Toxic America:
A 21st Century Analysis of American Culture

By Curtis R. Crim BA

ISBN: 978-0-615-44190-0

Printed in the United States of America

First Printing

This book is dedicated in memoriam to the vision of the Founding Fathers of the USA.

TABLE OF CONTENTS

Preface

Having reached the age of forty seven, I look back at the technology that was available when I was a child (we didn't even have a color TV set), compared to that which has evolved within our culture since.

Major breakthroughs in human technology have given rise to enormous change in human culture and physiology.

Among those major technological developments are the invention of the use of fire to cook food, that of the bow and arrow, and the industrial revolution. The computer revolution of the latter twentieth century has given rise to the 21st century Internet Revolution. Connecting all those computers together back in the 80's gave birth to an enormous cybernetic organism which now provides means of business, recreation, and the most advanced educational tool ever seen in human history.

The average American's life is so infused with and dependant upon computer processes that most of us don't even think about it anymore. Your PC, laptop or cellular computer are obvious, but your DVR also has an internal computer, as does your security system, your car, probably your watch, and your microwave oven and let's not forget about the computer that controls the automated factory where your Egg McMuffin was manufactured!

If all computers were to breakdown overnight, the average American citizen would have no way to get anywhere, feed his family, contact emergency services, or indulge in any form of recreation, or even communication! We, as a society, have become COMPLETELY dependant upon computers.

Our culture has also adopted the habit of accepting new technology without examining whether it might be accompanied by long-term detrimental effects.

Even the short-term adaptation to new technology can cause discomfort to some individuals. New technologies are immediately used for the exploitation of the average person. Telemarketers and spam emails are two examples of such exploitation.

The media itself, and I am referring to all forms of media, are used to manipulate people's minds on both the conscious and subconscious levels.

The effect of computers and modern day media upon humanity has both short and long-term consequences. It directly affects our culture and politics, while also perhaps leading humanity down a road that will lead to negative consequences genetically.

Our culture has become one that automatically accepts all technology as something that is good, where in fact we might be well advised to use caution with each and every introduction of new technology.

This unfortunate automatic acceptance of technology and media has made our culture, as a collective whole, vulnerable to control.

Introduction

Living in Southern Iowa, I am in the heartland of America. Massive corn production and a strong conservative rural culture characterize the world I see daily.

I, by contrast (which you will see in my writing) tend to be non-conformist and not terribly conservative.

In fact, I have striven for uniqueness my entire life. I consider myself outside of American or any other culture. Yet as an outsider, I have lived in the USA for almost fifty years, and have seen a disturbing trend back into the dangerous conformism that characterized the 1950's.

However, I am certain that I am not alone in experiencing the loss of rights in America. It has gotten so bad that not only corporations and the government take our rights away, but now even menial companies such as Internet Service Providers take rights from the average citizen (examples to follow in future chapters.)

I no longer feel comfortable sitting back and witnessing such a sad turn of events without saying something, and doing something about it.

One thing that breaks my heart about America is that we have all been collectively brainwashed into thinking, "there is nothing that I can do about it." The average person feels so intimidated by the gigantic Federal government, by Corporations, and by the ultra rich, that he automatically accepts that he has no real power in the world.

Believe me, the Ultra Rich Corporate Lords (URCL) want you to feel exactly like that. After all, we are brainwashed with the notion of freedom and democracy, where in reality we are slaves. In order for a slave to behave properly, he must feel utterly helpless, right? After all, the USA couldn't afford another Spartacus situation!

I feel strongly that we are not nearly as helpless as we are brainwashed and manipulated into thinking.

Perhaps it all comes down to attitude, and I am free if and only if I truly believe that I am. It appears as though the brainwashing itself is that which keeps me in "chains".

Therefore, I feel that a more in-depth look at brainwashing, propaganda, and manipulation of populations is in order to better understand current day culture in the United States of America.

Chapter 1:
Rome – the basis of American culture

To understand modern day Western Empire culture, the capital of which is the USA, I believe that it is important to look at ancient Roman culture, and consider it's evolution over the ages.

Some people think of propaganda as something that has come into being since the industrial revolution, and is used in modern culture. If one looks at history, clearly the practice of manipulating the minds of entire populations is no new concept.

It is likely that this practice goes back as far as the earliest civilizations. We as a culture are not comfortable discussing propaganda, because it is what our society uses to control us all. No one says it, but propaganda is a taboo subject in most cultures for this exact reason.

It is also for this reason that I am making propaganda one of the main themes of this book. What is propaganda, and how does it affect our lives?

Propaganda is one of the themes that most defines ancient Roman culture. A Roman named Octavius, grandson of Julius Cesar's sister, used propaganda to soundly defeat Mark Anthony and Cleopatra in the court of public opinion prior to winning his first military victory against them. Octavius, master of propaganda, later became Emperor of Rome, and took the name Augustus.

Most Americans think of a "barbarian" as an uncouth, crude, primitive, un-bathed, poorly dressed, violent and savage person usually armed with a big war axe. The fact is that the word "barbarian" comes from the Greek culture. It simply means "other" and refers to anyone of another empire. The ancient Greeks used this word to describe Persians and Romans.

After taking over the Greek Empire, the Romans adopted many of the Greek cultural traits. These traits include religion, politics, and architecture. The Romans also adopted the word "barbarian", and used it to refer to Empires outside the Roman Empire, such as the Gauls, the Daceans, the Goths and the Greeks.

However, the propaganda of the time brainwashed the Romans into thinking that barbarians were primitive and beastly, where in fact this is nowhere near the truth. The

Goths had science, architecture, mathematics and road building. They were an advanced civilization, as were the Greeks and the Daceans. In fact, most of these Empires were a good bit more advanced than the Romans of the time, but the population was forced to believe that they were animal-like people. This propaganda and spin on what a barbarian is still prevails in modern Western Culture.

It is important to note that lies told two thousand years ago are automatically assumed to be true today. If the population is still believing lies told eons ago, what of the lies and propaganda being spun in current American culture?

The widespread and liberal use of the media to enslave and exploit the general population is a cultural theme that has been inherited by the United States of America, and passed down from Rome itself.

One of the most destructive and dangerous forms of propaganda is censorship. Propaganda is not only spinning truth and non-truth to serve one's own purposes, but it is also the limitation of the truth that others are allowed to express and hear.

The Roman Emperor Constantine moved the seat of the empire's power from Rome, Italy to Constantinople, Turkey. He also chose Christianity as the official religion of the empire, a decision which has had dramatic consequences on the evolution of Western Culture.

In 325AD, Constantine called together the Council of Nicea (Nicaea), a meeting of all of the major Christian leaders. This council was the first to compile the Bible into a book. They decided which gospels to include, and which ones to omit. Through editing and censorship, a document based on propaganda was born: the Bible. Among the gospels that were censored and left out of the Bible entirely were the gospels according to Jesus Christ, Mary Magdalene, Thomas, Judas and Phillip.

That's a lot of gospels to leave out, and you would think that Christians would want to know what Jesus said first hand, not as witnessed by someone else!

Religion and mysticism after all, are our primary and most extensive source of propaganda and it always has been so, going back to the dawn of human culture.

It was important, in this first version of the actual Bible compiled into a single book, to tell a consistent "truth" to the Christian believers, NOT to tell them the actual historical truth as it was documented. It would upset too many Christians to know that Jesus was a murderer, so they left out all of the gospels about his childhood from about the age of 7 to about the age of 30.

The Bible is one of the most popular sources of propaganda of all times.

The Christian church quickly took over the Roman Empire, and although the Western Roman Empire fell long ago, much of its culture was preserved by the church and passed down through the ages via the royalty and culture of Western Europe, and eventually to the United States of America.

The Christian church has campaigned for centuries to crusade, but spin the reality of it differently in history. They have performed witch hunts, and incorporated it into western culture. They have created holidays (like Christmas and Halloween) just to de-emphasize pagan holidays that were popularly celebrated. The Christian church has manipulated and formed our culture over the ages to suit its own needs, and to favor the rich over the poor. They have also practiced the execution and torture of non-conformists in order to breed a more easily controlled population. The lack of intelligence widely seen in America these days is by design. Our population was created to be controlled and behave as society dictates.

The Roman Catholic Church continues to use Latin to this day, the official language of the Roman Empire.

The Western Empire centered in the USA still has Christianity as our official national religion. Although we claim the right to not be persecuted religiously, anyone who has had someone come to his door trying to convert him knows that it is common practice. I have one comment for the Jehovahs Witnesses: Jesus Christ and I have an understanding; we both think that Christians are evil.

The USA claims to be a democracy, but it is actually controlled by 1% of the population the Ultra Rich. The USA is clearly an empire, because it thrives on illegally taking over other nations.

Ancient Roman culture was obsessed with gladiatorial games. The daily spectacles in the arena included the killing of animals, people being fed to fearsome creatures, the execution of prisoners at noon, and of course gladiatorial fights.

The modern day cultural equivalent is America's obsession with sports such as football, etc. If a total retard has no idea what to have for a personality, he can always say "I like sports", and he will fit into American culture.

One characteristic that is fostered by sports is that of conformism. The Christian church is again guilty of pressuring individuals to become conformists, as is our culture in general. When faced with punishment, people frequently find it advantageous to not stand out. Our society rewards conformism, and punishes individuals who dare to think for themselves.

This arena-obsessed, propaganda-infused, conformist, aggressive, Christian Empire strongly resembles the Roman Empire itself.

In fact, I think of the American-based Western Empire as being the reincarnation and rebirth of the Western Roman Empire. We still even have crusades, which is a sad testament to what we have become!

Also note that the symbol of the Roman Legion was the Eagle Standard. The USA also uses an eagle as a national symbol. It is our national bird. I don't think that this is pure coincidence. Our culture is based upon ancient Roman culture. It is natural that we would share various cultural characteristics.

We now have some perspective on what propaganda is and how it has become a major part of modern culture in the United States of America. From the understanding that almost everything you see around you is propaganda on some level, let us examine modern Western Culture in more detail.

Chapter 2:
American Culture in 2011

Conformism

America is predominantly a conservative culture. Being liberal is generally looked down upon, and liberals are frequently persecuted by many factions in society including law enforcement.

This is one of the reasons that conservatives are also conformists. Each conservative knows the evil crimes he has inflicted upon innocent individuals just because they stood out in some way. Conservatives have no desire to show individuality, because they fear the punishment inherent in having an opinion.

Conservatives, such as Republicans, are therefore cowards. They are afraid of being individuals, and when persecuting minorities, they tend to attack in packs, so that their victims have no chance of defending themselves.

As much as it requires a great deal of bravery to express oneself as an individual in the USA, the act of being a conservative is downright cowardly.

In forty seven years of life as an American citizen, I have been persecuted and assaulted violently many times by white people for being black, Jewish, Hispanic, liberal, Native American, gay, a hippie, a terrorist, a cracker, a Democrat and a practitioner of the Transcendental Meditation technique, among other reasons. In reality, almost none of this actually applies to me, but when conservatives decide that they are going to persecute an individual or a group within society, the excuse they use is trivial. The reason that they assault minorities is that they are violent, cowardly, small-minded, repressed individuals with no concept of morality. The rest of it is just an excuse to ease their conscience for the crimes they perpetrate.

The fact is simply that conformism in and of itself is dangerous, if not inherently evil. I will cite the Holocaust during World War II as an example. The Nazis were not just socialists, they were fascists. This cultural attitude necessitates conformism, without which massive acts of genocide would not be possible. This is one reason why people who are violent and conformist by nature are often perceived as Nazis in general (law enforcement employees, for example.)

Historically speaking, every massive act of genocide has required conformism on the part of individuals. Extreme evil cannot exist without conformists. Conformism, again, is at least dangerous, if not outright evil.

It's not just that it is a lot of fun to conform to what everybody else is doing; it is also bloody awful to fail to conform, and end up being ostracized or exiled.

It is also to the advantage of the URCL's to rule over a conformist population, so the reinforcement to conform is poured into the mind of every American citizen every day.

To the non-conformist, the majority of the population appears to be totally retarded. This is actually not entirely true, as some conformists are only partially retarded, and too cowardly to stand out.

I will use Facebook as an example. This sad testament to the high school popularity contest as usual has all of the conformists mesmerized. When I heard that people actually keep track of the number of "friends" they have on Facebook, I knew that I would never participate in that web site. For the non-stupid, putting up any kind of internet profile is a really bad idea. It kind of spoils any attempt to live "off the grid", and fosters the kind of conformist brainwashing that leads people to disaster. I wonder if the conformists ever look at the non-conformist and wonder why they are not conforming. Are they even capable of generating a thought that was not fed to them by the culture or someone else?

I saw a comic on TV comment the other day that the Government no longer has to spend money to perform surveillance upon the population of America because everyone on Facebook is *giving up that information for **free!*** It was totally funny, but I think that it makes a serious point. The information put on Facebook is *not* information that you want strangers to know.

All of the information that you put on your Facebook profile is downloaded in an automated fashion, and then disseminated to various corporate, private, and government databases. The information you give up for free should be held as sacred and private.

Here is an example. A serial killer who is not a member of Facebook looks up the name of a girl he was in love with in high school on a search engine. One of the results points to her Facebook profile. He clicks on that link and sees a picture of her and the name of the town she lives in displayed. It is a tiny town in a western state. Now, all he has to do is go to the area and look around until she shows up. In a small town, it won't take long. Now he can do as he wishes and move on before anyone is the wiser. He didn't even have to become a member of Facebook in order to use it to stalk and kill an innocent person.

Here is the scary part: the information that she put on her Facebook profile is now available to nearly every sociopath, psychopath, and serial killer on the planet!

Facebook lures people in by appealing to their egotistical need for love and friendship. The actual outcome is that they make themselves vulnerable and helpless.

What, for instance, does the one lemming out of thousands think when he sees everyone around him jumping off the cliff to their deaths? Perhaps he looks down at the bottom of the cliff spattered with blood and corpses, and thinks that the majority of lemmings are pretty retarded. What of the conformist lemming? Does he see that one lemming unwilling to jump and think, "I wonder why that one…" <SPLAT>! The conformist lemmings thought is of course cut short.

My point is that after some psycho has hunted down and murdered your child is no time to consider whether Facebook might be a tool of evil, designed to brainwash people by getting them to conform. This sort of conformist recreational internet site is also a marvelous distraction from the corruption which runs rampant in American society. If early American colonists had not paid attention to the corruption, we might still be colonies of the British Empire!

The Loss of Rights

To examine any aspect of American culture, one must view it from two very different perspectives. One perspective is how an issue is "spun" in the media, and in popularly held belief. The other perspective is based on fact, scientific data, historical events and reality.

The way we in America are intended and expected to see any part of our society is based on propaganda and the spin that helps the individual be a good conformist and mindless slave. This perspective generally comes through the media.

As American citizens, we are expected to think that we live in a democratic society, and that we have freedom and liberty (which by the way mean exactly the same thing.)

There are in fact no degrees of freedom and liberty, but rather it is a binary situation. Either you are free, or you are not. If you freedom is limited in any way, you are a slave and not a free person.

We in America hold many rights to be self-evident, but what is actually in evidence is that our thoughts, feelings, behavior, and speech are all controlled by the Multinational Mega Corporations (MMC's) which are in control of our political system, our technology and our religions, if not our entire culture.

We are also brainwashed into thinking that it is okay to take our rights away. Ever since the year 2000, Americans have been conditioned to accept the loss of rights, usually based upon false pretenses. The Patriot Act of treason against the USA is a perfect example of this. I no longer have the right to not be tortured to death by rednecks without reason or cause because congress was bribed by the MMC's to defecate upon the Constitution of the United States of America. Their excuse is that there were enough congressmen who were retarded cowards that they were actually scared of terrorists! The only terrorists I saw were Bush and his band of cronies. Yet, I lost important rights because of corruption and cowardice. I have never met a terrorist who can stand up to this one, so the people who PRETEND to be my allies are far more dangerous, in my analysis. Here's a rule that might save your life: If someone tries to gain your trust, they are NOT to be trusted. It is logical, but beyond that "better safe than sorry". [Side note: all sayings are brainwashing.]

Most of the brainwashing spun in the USA is motivated by the desire of the MMC's and URCL's for profit. It suits their needs to lie to you and manipulate you, so they do it. If it is not legal, they just commit an act of treason and illegally bring influence to bear upon enough congressmen to get their desires met.

Because it suits their purposes, the URCL's impose the brainwashing of conformism upon you at all times through the media.

One example is the right to "freedom of speech". I consider this of utmost importance. Again, I do not wish to be a slave, and if I cannot control what I say, then I cannot control what others hear, and they then are controlled in terms of what they think.

The censorship of speech is essential to the enslavement of the American public. Congress was bribed by traitors to allow censors to edit the content seen and heard on radio and television broadcasts. The claim was that those forms of media are not covered by the constitutional right to freedom of speech. Since they were at the time, before the internet, the most powerful forms of mass communication, the most important forms of media to control to take away peoples' rights, and enslave them to the Ultra Rich 1%. This act of congress was treason against the USA. This is of no matter to the MMC's and the URCL's.

We American citizens have it drilled into our brains that we have freedom of speech and freedom of the press. With the domination of our media by censors, this is clearly nowhere near being true.

I do intend to go into more detail on the previous points further on in this book, but for now I just want to make the point that what we are told to believe is one thing, but what we actually get is something entirely different. We are told to not believe our eyes, but rather other people's lies.

Either you are a free person, or you are a slave. If you can say anything you want on television, then you are a free person. Since we cannot, we are slaves of society and of the Ultra Rich who control the media and our air waves.

Prior to the year 2000, most Americans believed in an ineffable quality in the constitution. We believed that there was something concrete there that could and would not be eroded. This general feeling has been obliterated by 10 years of treason and the trampling of the American slave class by the true criminals in America: the 1% - the aristocracy.

After the massive bribing of congress and the signing into act the treasonous Patriot Act (which is part of the destruction of the world trade towers scandal), the constitution no longer seemed to "hold water" for many Americans. This had a devastating effect on the collective consciousness of the USA. Again, the URCL's and MMC's involved in the attack showed no remorse. They appear to only care about profit, not human life.

Suddenly, an important part of the "heart" of the average American citizen had been carved away by the blade of corporate corruption. Our constitution was gone. I want to point out the irony that the safety of the Constitution of the United States of America at the time was placed in the not-so-capable hands of George W. Bush, the first illegal dictator of America in history.

We Americans have in the last 10 years become so accustomed to having our rights taken away, eroding slowly but surely on a daily basis, that we hardly notice or mention it.

Sure, anyone who has lived in my lifetime automatically accepts the federal or state government limiting or taking his rights away. Call it part of the cost of living in any particular country. It isn't fair, but it appears to be part of human society throughout history. There is nothing you can do about it. Or is that just brainwashing too?

We are also accustomed to having large, enormously wealthy corporations take our rights away. We accept that because we are programmed and brainwashed into believing that you cannot fight those who are rich and therefore powerful. This, it turns out, is not the case. It does appear to be true that powerful people are more difficult to fight, but not impossible to fight. This is an important distinction. Think of the ultra rich as a challenge. Perhaps they are more dangerous than the average person, but they are also more interesting targets.

Recently, a far more frightening threat has come to be. Internet Service Providers such as Mediacom and LISCO (Local Internet Service Provider) have decided to destroy the rights of the individual. I know it seems hard to believe.

My mom was switching from LISCO to Mediacom as her ISP, bundled with several other services. She called me and said that her old email address was not working. I asked her if she had cancelled it, and she said that she had not. I asked her if she gave the Mediacom guy permission to cancel her service for her, and she said that she had not.

I called LISCO and the person I talked to said that Mediacom had cancelled all of her services, not just the ones she wanted cancelled, and did it without her permission, and without even showing her the decency of informing her that they were doing it. I was surprised that LISCO would comply with an act that I considered illegal. After all, they had a contract with my mom that did not involve Mediacom in any way. Why would they disconnect ANY service without her authorization, let alone a service that she wanted to keep? How did these two ISP's get together and decide that it is okay to take people's rights away from them, without even consulting the law? They decided for themselves that it is okay to take the control of a service away from someone they have a contract with, and since the common person is a slave, it has no rights to worry about. Why is it assumed that the average person should have NO RIGHTS at all, and not even get an opinion on the matter? Through this kind of automatic brainwashing, American citizens have been made to accept on a subconscious level that they are worth nothing, and that they have no rights.

The reason for this is that it serves the purposes of the ultra rich 1% to impose upon Americans the idea that they have no rights. On paper, you might be a free person. In practice and in one's subconscious, one knows he is a slave.

This horrible state of affairs in the USA has come to be for a reason. The motivation to brainwash people into feeling as though they have no rights is profit ultimately.

In order to maximize profit, the MMC's and URCL's need a large slave class, which we call the poor. They have no rights as far as anyone can tell. It is okay to imprison them and force them to work as slaves, regardless of whether they are guilty of anything. It is even okay to kill them, if you feel like it. [It is a common practice throughout America for law enforcement employees to murder indigent individuals and dump their bodies. No one questions it.]

Although millions of innocent people are illegally imprisoned and forced to work without pay, which strangely looks like slavery, every person who is terrified into showing up in a cubicle daily because of unmanageable credit card debt is also a slave.

The simple and obvious fact is that the United States of America is an Empire ruled by the aristocracy, and is a democracy only in name, not in form. No empire in history has run effectively without a slave class, and in America the middle class is disappearing, as is the design of the 1%. We are all to be made into slaves, and one cannot allow a slave to have any rights!

Giving a slave rights only decreases the bottom line (profit, in vernacular.) As long as the corporations can enslave your mind, they can enslave you in actuality.

Don't let them do it. The loss of your mind is too great a loss; do not suffer the indignity of any society, even American society.

The Advertising Industry

Spam

One of the saddest examples of evil and corruption in the contorted corpse of what passes for American culture is the

Advertising industry. If ever there was an example of what Americans should be ashamed of, it is the American Advertising industry. These moralless monsters are used, metaphorically, as a weapon to "shoot" Americans in the head over and over again every day of their lives.

The Advertising industry is one of the most perfect examples of Pure Evil that you will ever see, hear, or otherwise experience.

The Advertising industry is based on lies, deception, manipulation, brainwashing, propaganda and subliminal attack.

Part of the evil of the Advertising industry is the brainwashing that it is okay to advertise to you. I do not feel that this is right. I think that advertising should be more voluntary and less forced on one than it is now. I also think that by intelligent and creative use of computers and computer systems, that we human beings can and will reclaim part of our dignity.

I would first like to address the spam industry. This includes any kind of unwanted email, such as spoofs that attempt to lure one into disclosing passwords and personal information.

The spam industry is just one branch of the Advertising industry. However, it does serve to victimize people, and over half of the spam must be from the pharmaceuticals industry alone. The pharmaceuticals industry is guilty of bribing senators and representatives alike on a massive scale. We are expected to accept the fact that any service we have can be used to victimize us and take away our rights.

I say that the average person, even if they are not rich (and most people are not), should have the right to say exactly who can contact them through their email service.

I am not talking about standard anti-spam services. Most of them either over-engineer their solution or make themselves part of the problem. Spamarrest, for instance, spams their customers heavily, and there is nothing that you can do about it. This is a classic mistake. It also assumes that Americans are so accustom to being victims that they won't notice it, or at least will accept it. This is implied brainwashing. You are not supposed to notice that a service you are paying to protect you from spam is ADDING TO THE PROBLEM!

Here is another point where LISCO's guilt is illustrated. I once had an email account with a local internet service provider company who call themselves LISCO. I was receiving spam emails from them, from a woman named Gillian Pierce. They were for services not offered in my area, so I asked her to stop emailing me. She continued to send me her spam emails, so I added her email address to the Blacklist of the spam control filter provided by LISCO to all its customers. Curiously, I continued to receive emails from Gillian Pierce (gpierce@lisco.com.) It turns out that she just bypasses the spam control service that they provide, and continues to spam LISCO customers, in spite of all desperate measures by her victims to defend themselves from her attack on their rights. Again, I should have the right to say who contacts me, and who does not.

What if a parent were to add Gillian Pierce's email address to the blacklist for his child's email account? That child would continue to be solicited by Gillian Pierce, because she bypasses the blacklist. Even parents cannot protect their children from Gillian Pierce. If LISCO is going to victimize children, then LISCO is evil. These people consider themselves enlightened, by the way, where they are actually the right hand of the devil, destroying rights and victimizing the innocent.

I think about the rich, and how they don't have to be victims of the Advertising industry. They can go to nice country clubs that have their own security services. They don't have to be annoyed by traveling salesmen as they play their leisurely game of golf. They have secretaries, personal assistants and security systems in their homes. Is there a reason why the poor (the majority) of Americans cannot have the same protection and live the same dignified lives as the rich?

Historically speaking, computers were first invented to make people's lives better. Quite quickly, they were perverted to assault people instead. The spam industry is a perfect example of this. Automated software is now used to exploit Americans as much as it possibly can.

As an engineer, I see that it should be just as possible to use computer systems to defend the innocent, and restore their rights, rather than taking them away the way the oil and pharmaceuticals industries do.

Rather than just complaining about the state of the world, it seems more productive to present solutions that can actually help restore rights.

If one were just to create a web based service that serves the same function as the security at a country club, then users could have complete control over their email.

Most email spam filters make the mistake of over engineering and making too many lists. Why maintain a blacklist when the spammers will always think of a new way to spell the word “software”? I was brainstorming an implementation for a block-by-domain feature (another feature rarely offered by spam control services) when I got the idea for a white-list-only filter. Rather than making me jump through hoops trying to say who CANNOT email me, why not just block *all* email addresses, and only let in the ones that I specify?

This gave birth to the idea of Schpleee Technologies’ Green List Technology. I also call this Email Sender Management, because that is what you are being given the power to do, manage senders, not just individual emails. We will see that it can be applied to any service, whether free or one that you pay for.

If you are lucky enough to already be rich, and I mean like, super rich, you would probably have security handling who can enter a party you are throwing. The security company is very competent, and asks you for a list of whom they will be allowing in. If someone is not on the list, they are not getting in, and that is the end of that.

We are now going to implement that security guard as software standing as sentinel watching over your email. You will get to provide a “Green List”. If the email address is not on the list, it is permanently deleted. I cannot go into too much detail in this book, because of space considerations. If you want more details, I suggest that you search the web for technical specifications by Schpleee Technologies Inc. and Curtis R. Crim. I plan to implement my green list technology as actual internet services which are not yet up and running (and if they are, yea!), but will be when I acquire the funding to get them developed. These web sites will include EmailParentalControls.com and IroncladSpamblocker.com, both of which will give members control over their email service(s) as they have never had before.

Most corporations attempt to maximize their own profit as they destroy the rights of American citizens. Schpleee Technologies Inc is attempting to make the world a better place by restoring rights.

Telephone Service Abuse

Do you ever wonder why you pay for telephone service, and yet people can harass you with prank calls, political calls, surveys and telemarketing? After all this is YOUR service. You pay for it, and it is between you and the phone company.

Last century, it was desirable to allow anyone to contact you by phone. But now that computers are being used to exploit the public, one's phone service has become a liability.

I was shocked when I realized that I don't only have to deal with telemarketers, but now they have computers that only connect me with a human after an actual person's voice is detected. It is okay to waste my time to advertise to me, but NOT okay to waste the time of their employees! This is outrageous.

After all, all I have between the moment I was born and the moment I die is my time – my seconds in life. Most people like to spend time with their family, or perhaps sell them time to an employer when they are paid to not be with the ones they love. When someone calls you on the phone without your permission, they are taking away the moments of your life that are precious and dear to you.

How about the person who wants you to take a survey? Are they paying you for your time? Data has become a very valuable asset. Companies used to have to pay other companies that correlate marketing demographics huge amounts of money for this information. Now, it has become a common practice for companies to go directly to the customer to ask them to take a survey through the mail, the telephone or on-line. They rarely offer to pay you for your time. They don't have permission to ask in the first place, but being a slave you are now accustomed to being exploited in such a way.

However, the engineer in me again reminds me that if a computer system can be used to exploit people and take their rights away, one can also be used to give a person control over their telephone service.

Therefore, I am introducing the Schpleee Technologies Inc. Caller Management System. Like the Email Sender Management system, this device will be based on Green List Technology. This system will use the Caller ID data and allow the user to specify a list of whom they want to be called by. If a caller is not on your list of approved callers, then the phone will not even ring. If you ban a caller who is harassing you, then their number will never again show up on the list of incoming calls. The point here is merely to put complete control into the hands of the receiver of the service.

Some phone services actually offer a limited form of this service, but it is awkward and limits users to a list of 12 numbers that they can ban, and 12 numbers that they can green-list. The Schpleee Technologies Inc Caller Management System will allow up to 100 numbers that a person can put on the green list. A version with more memory will be released at a later date.

Once a user is protected with this caller management system, they will never get a prank or crank phone call, someone from a charity, someone wanting them to take a survey, or vote for their candidate, or even a wrong number! For the first time, people will be able to enjoy their phone service without the liability of being exploited for their time by the American Advertising industry!

The Schpleee Technologies Inc Caller Management System is not yet available for retail sale; however look for my technical specification, which will be made available on line.

I know that it is insane to sell technical specs on-line, but I am so desperate to personally enjoy the technology that I am proposing, that I will thank the gods if anyone at all will implement it. If not, it is my hope to offer this system in the retail marketplace in my lifetime. This is also my hope for my email Sender Management System. I plan to release it as soon as possible; I have been through many spam filter services, and have yet to find ONE that handles it correctly.

Television Advertising

It has been pointed out to me that the most retarded people actually enjoy television advertisements. I am sad to say that I believe that this is true, but I also believe that most people do not like ads on television.

One of the coolest things about TV advertisements is that you are brainwashed by multiple levels of bullshit at the same time. The fact that they put a commercial in your face is bullshit, even before they start lying to you. By running commercials, they brainwash you to believe that it is acceptable for them to advertise to you, and it is not. Finally, after filling your head with lies and facets of culture-specific bullshit, they finally succeed at manipulating you into buying their filthy poison, and that is bullshit too. When you see a television commercial, you are experiencing at least four different levels and kinds of bullshit simultaneously.

I could bore you with a detailed dissertation as to how evil, effective, and powerful the advertisement industry is, but it would be a reiteration of chapters in my last two books.

Instead, I will point out that TiVo once came up with automated advertisement screening and that they were destroyed in a court of law and enjoined from releasing this technology. This shows that it is not only possible, but also that there is a market for such technology.

Further, DVR technology uses hard drive file storage space, so it will be advantageous to the customer to conserve file storage space. It is tiring to be forced to record hundreds of commercials, and then be forced to delete programs you want to keep in order to record another show.

Digital television signals have bandwidth not displayed on the user's television screen. Additional program data is transmitted in this bandwidth that allows DVR consoles to know when shows begin and end, and what their titles are, among other information. These signals also contain tags that delimit the beginning and end of commercial breaks. This information can be used to heuristically detect and eliminate commercials from a video data stream.

Television advertisements are used to indirectly victimize parents by manipulating their kids with brainwashing. A service that would allow parents to have control over not only the programming their children can view, but the commercial advertisements as well, would give them the control that they need and deserve, but that which is unknown in American society. It is too important to control people's minds for the profit of the corporations. Better to brainwash people into poisoning their children to death with soda pop and junk food!

It occurs to me that there is a natural and obvious combination that would benefit everyone with the marriage of internet and TV programming technology.

Too many forms of technology are pushing TV programming into obscurity. On the other hand, the internet offers the convenience that most people demand these days. My idea is to offer a full TV programming package service with DVR functions and unlimited storage space worldwide. Customers will no longer have to delete one program in order to record another. They will be required to pay a monthly fee for file storage space used. The convenience is that their DVR service will be available from any mobile device connected to the internet, and they will be able to play back programs on any device no matter where they are, as long as they have an internet connection.

This will bring TV programming back into mainstream American culture, and give the customer the best of both worlds, without requiring yet another console system in their living room.

Obviously, it would be crazy in terms of the use of resources to allow customers to use up HD space on commercials, so removing them from the video stream would be both good judgment and common sense.

As a side effect, the customer would be protected by the onslaught of propaganda and brainwashing inherent in commercial advertisements. This service would simultaneously give Americans added recreational benefit, as well as give them the opportunity to not be victimized by the TV advertisement industry.

Here is an example of advertising strategy that offends me: I bought some Jenny-O turkey bacon for my chickens (I prefer the Luis Rich/Oscar Mayer for human consumption), and there was an advertisement on the package for a lousy TV program. How far has our culture deteriorated when TV networks advertise on food? What freaking genius in the advertisement industry thought this up in the first place? If that individual reads this, I beg him to commit suicide before he harms society even more.

On the note of NBC, I want to point out that they have displayed some of the worst judgment in human history. When the original Star Trek was cancelled mid-third season was one of the stupidest things ever done, even in term of the production of profit. The judgment of the administration of any corporation is garbage. This is even truer of the television networks. I understand that Fox and NBC are particularly retarded, but who can really tell the difference?

One point that I want to make that is extremely well illustrated by television advertisement is the exploitation of human sexuality in advertisement. Now, I don't want to appear to be too conservative or proper. However, I do want to point out that the advertisement industry has never failed to use human sexuality to exploit the potential customer. The advertisement industry knows that magazines with sexy women wearing next to nothing on the cover sell better than magazines that display some other form of imagery. It is not the model or celebrity on the cover who is being exploited so much (and they are), but it is the customer who they are attempting to stimulate into buying the magazine who is being exploited.

I personally think that the level of modesty that some extreme Muslim cultures practice is overboard, but so is the extreme flaunting of sexuality of American culture. Perhaps the best solution is somewhere in the middle.

One thing that I think escapes most people is that television advertisement is not automatically benign. TV ads are so integrated into American culture that we don't even really notice them anymore. A friend commented to me, "They gotta' have their commercials!" We are brainwashed into thinking that it is alright to advertise to us. We are programmed to assume this.

Even the fact that they put TV commercials on the air implies that it is okay for them to do so. Don't accept it; question it. Consider the possibility that it is not okay, and that it should be legislated otherwise for moral reasons.

My main point here is that people should (and must) have the right to control the advertisements that are allowed to potentially pollute their consciousness. Advertisement, by its nature, removes rights from the individual. Commercial advertisement elimination technology, by its nature, restores rights to the individual.

Television advertisement also employs techniques that subliminally affect the watcher in such a way as to inflict mental retardation. Watching commercial television advertisement reduces one's higher brain functions, ability to learn, and state of consciousness. Television advertising takes away your rights and retards your brain function at the same time.

The American Advertising industry became powerful and thrived through the medium of television. They have now found a way to infiltrate every form of media that you can possibly imagine, including billboards, radio, TV, theatrical movies (product placement and ads before the show), the internet, the email, the telephone, magazines, even on labels on the food we eat. They literally are in every aspect of our lives. This is part of the threat of the Advertising industry. They are nearly impossible to avoid in modern culture, unless you live in a forest, that is.

The only moral solution is for DVR console and service technologies to offer the user the ability to remove commercials from the programming viewed by their family. I believe that the experience of being human can be enhanced by reducing the amount of advertisement we are exposed to.

Propaganda

Propaganda, also known as bullshit, is one of the major themes in America, as our media is constantly used to brainwash and manipulate us on a subconscious level.

I have stated that the practice of using propaganda to control the population of an empire goes back at least as far as the Roman Empire, the template for the Western Empire controlled by the USA.

Before proceeding, I think that I should define the word bullshit, which I am using as a synonym for propaganda, among other things.

The term bullshit is widely used in the USA to refer to something which is false or untrue. It is secondarily used to refer to a situation that is unjust or unfair.

I am using both of these definitions, and expanding the term widely for use in this book.

When I use the word "bullshit" it refers to, among other things, something that is untrue or a lie, something that is wrong, immoral, unfair, or unjust. This includes situations that are somehow painful, distressing, or uncomfortable. Example: Some guy gets injured while skiing. He might say, "Man, my broken leg really hurts; what bullshit!"

Further, it also refers to any spin on something, even the truth or an assertion using the truth. Any "spin" on true facts is still bullshit; thus, propaganda can be utterly true and bullshit simultaneously. "Bullshit" refers to any perspective that is biased by culture in any way, including (and especially) religion. Any perspective of any society is automatically bullshit. Bullshit doesn't have to be untrue or even bad. Bullshit is merely the way people look at the world. If people tell you how they see something, then it is bullshit. It's just *their* bullshit.

I am also loosely using the word bullshit to refer to any aspect of human culture, society, or civilization. This includes all politics, religion, mysticism, and popular belief. Bullshit is exactly what all human culture and society spins, just as a spider spins its web. In fact, any opinion from anyone's perspective can loosely be seen as bullshit, regardless of whether it is true.

I find that people are generally not comfortable with reality, and actually prefer some kind of bullshit to the truth. I have a saying: "Denial of reality is always going to win out over ACTUAL reality." People tend to favor their feelings over actual fact. The "bullshit" proves a comfort zone.

When people are born into a family, they are quickly exposed to one of the primary levels of bullshit in our culture: Familial Bullshit. One of the first things that people have to deal with in life is crap coming out of other family members. I have a saying about this situation: "I grew up in a family where being able to understand word salad was a requirement for survival." My sister is a master of "word salad". This is the practice of saying things that are designed to elicit an emotional response, while not making any actual sense.

In the next section, I will go into more depth as to the nature of the emergence of societal bullshit from its source in the Field of Pure Bullshit. This is part of my ultimate explanation about the nature of reality.

I want to point out the inherent brainwashing in the American primary educational system. I have pointed out that censorship is part of any campaign of propaganda. For a long time, it has been a tradition to not tell kids the truth as to the nature of reality. It is not just what the students are taught, but also what material is left out of the curriculum, that illustrates that the violation of the separation of church and state has long been a tradition in America.

Because brainwashing children has a higher priority than educating them, many historical points are "smoothed over" in American classrooms.

I know of no public education system that tells in detail the truth about the atrocities committed against the Native Americans. The public educational system conveniently skips over a majority of the truth about what happened during the crusades because the truth is so ugly and so awful that we don't encourage knowledge of it in our culture. The public educational system calls George W. Bush the X president of the united state, where in fact we had no president between Clinton and Obama. GWB was an illegal dictator and that is what the historical record shows. The American system is no system of education at all. It is a system that prepares children for college (where real education begins) and brainwashes them with propaganda so that they will make good slaves for the corporations.

I name the system we have in America our System of Advanced Slave Training. After all, they have no intention of educating the youth of America, and the corporations need slaves, not independent human beings.

In fact, it is possible for children to understand, process, and remember actual education on the college level as early as 3rd grade, but education gets no priority in America.

Not only do we not intend to educate our young, the corporations actually profit from keeping them uneducated. Remember, stupid slaves are more easily manipulated and therefore controlled than educated ones. Education leads to freedom and liberty, and ignorance leads to slavery. The MMC's and URCL's want to keep America enslaved; therefore actual education of the public is not acceptable to their business plan.

In fact, knowledge in general is not good for business. It is well documented that the stupider the American public is the more profit is generated for the corporations.

Because of this, it is not advantageous to corporate America to allow the truth to be told even in the news. The brainwashing that is generally accepted is that the news always tells the truth and the whole truth, like a court of law. The reality is that the exact opposite is true.

The news television stations begin with the selection of what they are going to report. This is where the propaganda and brainwashing begins. If the corporations do NOT want you to know about something, then it will not be on the news. If an article is on the news, then it is intentional, and is by definition some kind of bullshit. Lies of omission are among the forms of subterfuge that they practice. They censor and edit the news, they decide what will be reported and what will not (which is the majority), and then they spin what is left in an attempt to brainwash anyone who is a viewer and is retarded at the same time. The news presents a highly-spun highly-edited and tiny fraction of the truth. Some of the time, it is outright bullshit, like the coverage of the aftermath of Katrina disaster. The storm was not so hard on the survivors as the dictator. If you watch the television news broadcasts and fail to understand that most of what you see and hear is bullshit, then you are retarded.

Like my mother.

The entertainment industry is also guilty of spinning bullshit in massive quantities. "Product placement" advertising is one example of the brainwashing that they put into movies and is difficult to screen out of one's consciousness.

Value systems are also programmed into movies that influence how we see the world. I saw a movie the other day that had a timeline that bounced back and forth between the 1940's and the early 2000's. In each timeline, there was a woman involved in the cooking of food. They were both married to husbands who loved them very much. In both time frames, the husband was worthless in the kitchen, and was never shown doing something as simple as getting himself a glass of water. This is a truly antiquated stereotype of the average American household. I understand that it was accurate to depict the 1940's culture in this way, but it seemed inappropriate to bring that same stereotype into the culture of the early 21st century. The creators of that movie were clearly imposing their old fashioned stereotypes on current day culture, and it just made them seem ignorant and desperately out of touch with the world of today.

My point here is just to illustrate that any movie has its own spin on how to view the culture, and in each case it is bullshit.

Political systems are also full of bullshit. Look at the various political systems world-wide and throughout history. These include monarchies, communism, empires, socialism, dictatorships, "capitalism" (not a real system of politics), and "democracy" (also ultimately fictitious.)

Regardless of the official national system of politics, there is one common underlying theme which is the fact that the aristocracy ultimately always controls the political system.

In monarchies, dictatorships and empires, one ruler or ruling family owns everything and dominates the entire society. Even in socialism and communism, the bottom line comes down to human greed and the most powerful and rich aristocracy still ultimately controls the "political" system.

The problem with democracy is that it will always be subverted by greed. The USA is a capitalistic society, the operative root word being "capital", or money. Any system based on money will always be infected with the cancer of greed. When you have evil ultra rich aristocracy and corrupt congressmen, you will have only a façade of democracy.

The term "democracy" is used commonly in America to brainwash the idiots, bread in

Western and Northern Europe and infecting the lands of North America.

Going back to Roman history, we find that they also had corrupt politicians. Greed and corruption are not new concepts to our Western democratic system of government.

The word "politics" is defined as a struggle between classes. In America, we are brainwashed into thinking that is a struggle between the left and the right, liberal against conservative. In fact, this is more brainwashing by the aristocracy to keep the masses fighting with themselves, while the ultra rich 1% victimizes us all.

I was appalled to see Exxon/Mobile advertising on the television and on the internet during the last presidential election. I am convinced that they had a hand in forcing a dictator illegally into the Whitehouse in the year 2000. They were using an apparently non-rigged election to get the American public's confidence in the very people who had just rigged the last two elections. One can see the corruption of our political system mixing with the corruption of the Advertisement industry, and the corruption of the MMC's and URCL's.

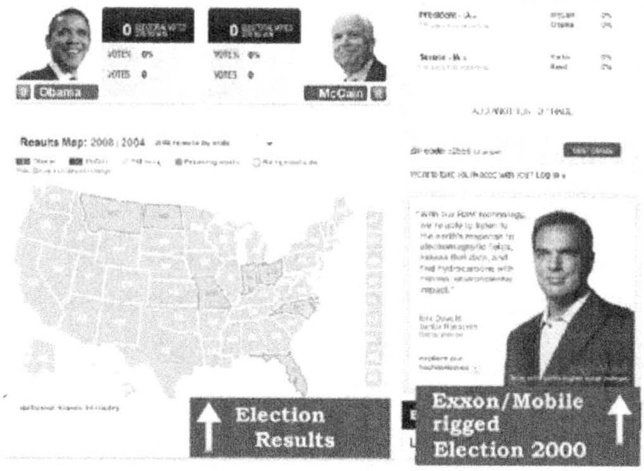

Figure 1: Exxon/Mobile Bullshit

The bottom line is that all political systems are bullshit all of the time.

Being a religious nation, it is necessary to consider the level of religious propaganda, brainwashing and bullshit.

I have defined bullshit as meaning, among other things, any cultural spin on any subject. In no part of culture will you find bullshit that is as intense and absurd as religion. All forms of mysticism and spirituality go into this same category. Since these belief systems are meant to explain things in the universe that humans do not understand, they by definition and necessity are purely bullshit.

In fact, the word, "Holy" translates directly as "Bullshit". The Holy Bible is the Bullshit

Propaganda. The Holy Tradition is actually the Bullshit Tradition.

Religions have been the motivation (and often funding) for religious wars such as Jihads and Crusades. They have been responsible for inquisitions, witch hunts, and genocides. They have masqueraded as spiritual guides, while actually being motivated by the acquisition of land, slaves, money, and power. Religions are the ultimate bullshitters.

Religions, in fact, are so good at bullshitting that they even brainwash those who don't believe in religion at all! If you ask anyone on the street what atheism means, he will frequently reply that an atheist is "someone who doesn't believe in God." Do you see the inherent bullshit and brainwashing? Why don't people say that an atheist doesn't believe in "the Gods" or "in any Gods"? It is because in our culture the bullshit monotheistic brainwashing of the Christians causes even non-believers to accept that there is only one God, even if they *don't* believe in God! As a good atheist, I think that a person should take the time to not believe in all of the other Gods in all religions, polytheistic religions as well. If you really don't want to believe in any Gods, then you should put attention on not believing in the Old Norse Gods, the Hindu

Gods, the Native American Indian Gods, the Druidic Gods, the Wikken Gods, the old Greek and Roman Gods, as well as the obvious Gods from monotheistic religions, like Buddha, Yahweh, Jehovah, and Allah. Anything less is a waste of time to a true atheist.

In society, we also have to deal with social levels of bullshit. Gossip is an excellent example of that. People who have no lives and no intelligence enjoy gossiping about other people's lives. In each instance of a story, it becomes less factual and more exaggerated until in the end it no longer resembles the truth. However, if it is good gossip it will be believed and continue to be passed on, and increasingly distorted.

My rule is to assume that anything anyone tells me is total bullshit, until I can verify what kind of bullshit it is, and where it came from, and how badly it stinks.

The picture of American society that is shaping up is that it is layer upon layer of bullshit, going back as far as Rome, gathering bullshit from our founding fathers, and stacking upon it layers of political, social, cultural, and religious bullshit.

If you belong to a cult or spiritual movement of some kind, then they also will have their whole rhetoric laced with bullshit.

My sister has lived with the Transcendental Movement her entire adult life, and moved onto the Fairfield Capital of the Age of Enlightenment right from our parent's household. In her opinion, you are either with the TM movement or you are not. There is no in between ground. If you are not practicing TM, you are going to hell.

I also lived on Maharishi International University (now MUM) campus and got my degree in computer science. While there, I was able to enjoy a whole entire new level of bullshit on top of the many layers of bullshit that most members of American society have available to them. I was in bullshit heaven!

I also learned of higher states of consciousness, and that it is possible to use one's consciousness to contact the source of the bullshit universe, the source of all bullshit, known as the Field of Pure Bullshit.

Enlightened bullshit can be achieved when a person's consciousness evolves to become one with the Universe, also known as UB, or Ultimate Bullshit consciousness.

I also leaned about how the universe of bullshit around us is created from the Field of Pure Bullshit.

To summarize, the self-interacting dynamics of the Bullshitter (Rishi), the process of Bullshitting (Devata) and the Bullshit (Chandas) create a resonance that causes the emergence of space, time, energy and matter fields that make up the bullshit universe around us (see Figure 2.) Since there wasn't a Unified Field chart for the discipline of bullshitting, I created one:

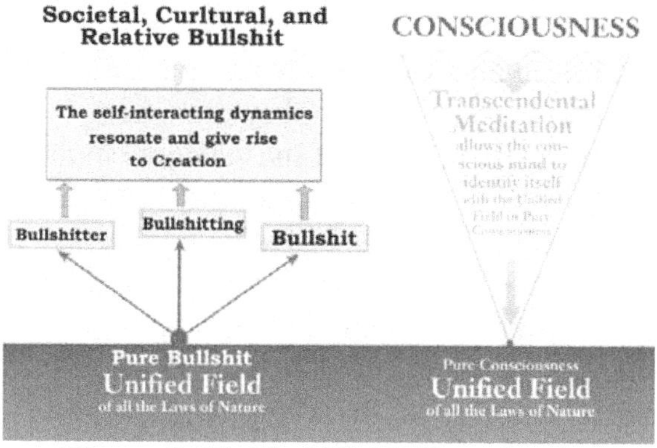

Figure 2: Unified Field Chart for the Discipline of Bullshitting

The source of the universe is the Field of Pure Bullshit, where the source of cultural bullshit of all kinds is the ultra rich 1%, the aristocracy of society. The reason that the aristocracy brainwash the poor masses with bullshit is that they want to control their slave population. The most effective means of controlling a slave is to control its mind.

Because any perspective is bullshit, and because consciousness cannot be separated from perspective, consciousness by its nature spins bullshit. The very definition of "bullshit" is that it is the perspective of consciousness. Obviously, if you can experience Pure Consciousness, you are also experiencing the Field of Pure Bullshit.

One of the interesting things about MUM campus and the TM movement is that they are so VERY full of bullshit! For instance, one is told about Guru Dev and the Holy Tradition. They have no way of proving scientifically that any of these people in the "Holy Tradition" were actually enlightened. Their claim about these "Holy" people is not scientific in any way. It is merely based upon belief. It is totally bullshit, because whether these people were enlightened is merely a leap of faith, and is handed down from a different culture, and a different time. Just because someone is painted with a halo over his head doesn't mean that he was actually enlightened in life. It is just an artistic embellishment.

Being egotistical and being "enlightened" seem to go hand in hand. I was sickened by how many people on campus have some fucking ego trip over their "enlightenment". Many people would look down on me for "not being Auyervedic" enough. I rode a motorcycle, wore a leather coat, and wore blue jeans, all of which was considered "unenlightened". Some of these movement true believer retards would carry their thermos of hot water every day, even in the summer. They would go outside wearing a great coat and with a parasol, even in temperatures over eighty degrees. They would do all of this because it was considered Auyervedic, yet they would also look down upon the student who did NOT carry a thermos of hot water.

The townspeople of Fairfield are rather conservative, and don't like people who are different. They call the meditators "Gurus", which is intended to be insulting. I personally understand why they get tired of being treated as though they don't exist or count at all, because they are obviously not meditators. The Gurus treat the townies as though they are animals with no feelings. Since they are "enlightened" they feel as though it is okay to treat other people like trash, especially non-meditators.

This is a curious development: If you consider yourself enlightened, it is okay to treat other people like shit – especially if they don't meditate, or you consider yourself more Auyervedic or more enlightened. Now, I would think that if someone is *truly* enlightened, he would treat other people with kindness and respect. Instead, it is just another excuse for abuse and egotism. I call the Transcendental Meditation a Bullshit Movement (also known as the TM BM.)

As for these "enlightened" idiots, I have seen some of these people using a cellular phone while driving. Attempting to murder children is hardly enlightened; they can leave that bullshit to the uneducated townies. If a person has been in the Dome for decades and wants to think that they are "enlightened", then they can *put down the fucking cell phone and drive already!"*

The movement also has what they call "Behavioral Rasayana". This is a list of ways in which people should behave. It includes sayings like "Speak the truth, but speak it sweetly." These might be noble sentiments, but any attempt to control a person's behavior is also mind control and therefore slavery.

Censorship and Mind Control

The lack of rights is what defines a slave. If an individual cannot control his own mind, then he has no rights, and is therefore a slave. The whole purpose of censorship is the process of possessing the minds of the masses.

In America, we have the bullshit of "freedom of speech" drilled into our heads regularly. In fact, we have nothing of the kind.

Many statements are considered treason, and therefore are prohibited. Corporations are striving to censor and limit freedom on the internet, and of course we are denied free speech on the television and the radio. Also, many fundamentalist assholes will take objection to profane language being used in public. Who the fuck told them that they can control my speech and my mind just because they had the bad judgment to have kids? I say fuck them and fuck their kids! These cowardly conservative assholes just don't want their five year old to call them a cocksucker. I say that anyone who attempts public censorship of strangers deserves to have a five year old kid call them a cocksucker! I never would object to someone using profanity in front of my kids, but I know that my kids are smarter than they are, and will not be harmed by hearing foul language.

The real reason behind this stigma against children swearing is because when a fundamentalist hears a child say "fuck" or "cock-sucker", subconsciously (or perhaps sometimes *not* so subconsciously) they think about fucking that child or having it suck his cock. It is again the disturbed psyche of the conservative repressed Christian censor that motivates him to try to manipulate the behavior of individuals in society.

Here are a couple of examples as to how bad censorship has gotten in America in the last 10 years: I was watching a comedy on comedy central and they actually bleeped the word "wet"! The context was that a woman was saying that something made her wet. Now, surely this has clear sexual connotations, but is still relatively harmless. The censor I am sure would argue that it is not the word, but the way it was used that was objectionable. The reality is that this fundamentalist repressed asshole was made sexually excited by this use of the word "wet", and therefore decided that it is not appropriate. In other words, it was this conservative piece of shit's own perverted mind that made the interpretation bawdy, and served as the excuse to inflict mind control on the viewing public. What a sad state of affairs!

These conservative fundamentalist bastards have also gotten entire shows cancelled, such as Sarah Silverman for being too raunchy, and Caprica, for portraying monotheists as terrorists, which they are by the way. They have gotten entire episodes of South Park censored, like 200 and 201, even stopped from streaming on the internet! These censors are completely out of control and must somehow be stopped!

A few nights ago I was attempting to watch a stand-up comic on comedy central after midnight. I would think that they would ease up on the censorship that late at night, but no dice. There were so many long and short bleeps that it sounded as through I was listening to Morse-fucking-code!
This form of mind-control, “censorship”, increased dramatically in scope and intensity during the Bush dictatorship.

I find it offensive that most people subscribe to the brainwashing that parents know better than their children and should be obeyed. I know it says that in the Bible too, which defines it as bullshit. Although lacking the experience of their parents, I find that many kids are superior in intelligence and judgment.

Much of the bullshit we accept in America is implied, and not stated outright. I would like to use the American Film Association and the American Film Ratings Board as examples of implied bullshit. Since these people are imposing their own feelings about films on the public, it is by definition bullshit, but we accept that. What I find funny is when you see these ratings displayed at the beginning of previews for upcoming movies. They say stuff like, "The following preview has been approved for appropriate audiences." This is totally retarded. There are no appropriate audiences. I find that audiences are exceptionally inappropriate, so you might as well let them see the preview anyway. The real bullshit is that they even have the audacity to rate movies and previews in the first place. Just by placing their rating at the beginning implies that it is okay for them to rate movies, where in fact it is not. It is both censorship and brainwashing at the same time.

In fact, the censorship of the TV and radio airwaves is a direct violation of our constitutional right to freedom of speech. The conservative fundamentalists were able to push legislation through congress to obliterate this right because our government is totally corrupt, and the ultra rich want to control what people can say and hear.

One finds oneself wondering WHY these asshole censors want to control what some people can say, and what other's can hear. Why is destroying freedom of speech so important that they were willing to commit treason?

The answer is mind control. If you can control what people can say and what people can hear, then you can also control what people can think.

The goal of the Ultra Rich Corporate Lords is to control the minds of their slave population. If you can control slaves' minds, then you can control their behavior, despite any "rights" they might have on paper. The goal of the ultra rich is to control the minds of the population in order to keep them in a perpetual state of slavery.

The saddest part of this is the hypocrisy of spreading the propaganda and brainwashing that we have freedom of speech, where in fact the aristocracy has taken our freedom of speech away from us as a means of keeping us enslaved.

Greed: America's Highest Value

At the risk of being a hypocrite, I would like to refer to the section of the Bible that discusses the seven deadly sins.

Even people who don't believe in the bible can agree that there are at least a few good notions in it. I also want to appeal to the people who DO believe in the Bible, because I do agree with its philosophy on the following point.

The Christian Holy Bible states clearly that greed is one of the deadly sins. Assuming that this is true, one might ask who this sin is deadly to.

The answer has been supplied by the Multinational Mega Corporations. Their greed is destroying the entire troposphere. The answer is that greed will lead to the death of all humanity.

Greed is deadly and sinful because it is not a moral value. It is a desire placed into humankind's collective heart by the embodiment of evil.

The problem with a capitalistic society is that it is far more vulnerable to the evil of greed that other non-profit-based societies. Societies that highly value spiritualism, for instance, are likely to be far less corrupt than a society that is based on money and the attainment of material prosperity.

Because America is based on valuing money above moral rectitude and decency, our system of society cannot be separated from greed, which we know to be sinful and evil.

As long as America remains a capitalistic society, it cannot be separated from the evil of greed. By its very nature, the USA is and probably will always remain evil. As long as the ultra rich are allowed to live, corruption will always be part of our political system.

Rampant Corruption in American Society

For argument's sake, let's assume that greed is causing corruption on all levels of American society. Let us now examine some of these forms of corruption on a more detailed level.

One of the biggest scams in America is the insurance industry. In fact, it not only makes huge profits for the corporations, but it does almost nothing to help its customers. It mostly serves to deceive and victimize people.

If an insured person never makes a claim, the insurance industry is happy to take his money until he dies, and never give anything in return. This is a very convenient business to be in!

On the other hand, if a customer actually needs to make a claim for any reason, then his insurance company will simply bend him over and ass-fuck him as hard as it can. You actually pay for this service.

Their first step is to find some lame excuse as to why they don't have to pay the claim. If they choose not to pay, there is nothing that one can do about it. The insurance companies are also amazingly dishonest, and lie constantly. The most evil of the kinds of insurance is the medical insurance industry. The people in the medical insurance industry continually try to screw over the doctors, who in turn try to screw over the insurance companies. The sad outcome is that the patient is stuck in the middle and usually gets fucked from both ends – by the insurance company and by the doctor as well. The only solution to this problem is to get rid of the medical insurance industry. After all humanity survived for thousands of years without them, I am confident that can do it again.

The pharmaceuticals industry is also reeking of corruption. Sure, they pump their Viagra and nasty little concoctions over the internet, on the television, and in the email. I think that Viagra is bullshit, because if a man uses it, he loses the ability to gain an erection without the use of Viagra, and therefore develops a dependency upon it. Side effects from modern pharmaceuticals kill people all of the time. The pharmaceuticals corporations are nothing less than drug dealers trying to keep America dependent upon their vile potions.

But that is not my real problem with the pharmaceuticals industry. They have time and time again illegally brought pressure to bear upon the Congress of the United States. They employ measures of treason to promote their own products while suppressing the availability of superior competing products. As usual, our legislature is being used to take our rights away for the sake of the profit of the corporations.

One might wonder who stands by an industry as evil and corrupt as the pharmaceuticals industry. The answer is the chemical industry.

The chemical industry supplies a huge number of products that we use in America, including plastics and polymers. Without plastic, our culture would be very different than the one we see around us in 2011. In fact, a majority of the chemicals and pharmaceuticals consumed in the USA are refined from crude oil. The major oil corporations actually own the chemical and pharmaceutical companies, just as they own the soda pop companies, and the fast food chains.

We now see that it is actually the Oil Corporations who are behind the corruption of the corporations which they own. The oil corporations have also had a long history of bribing congressmen and illegally bringing pressure to bear upon the federal government, with the motivation of maximizing profits while enslaving and victimizing the American public.

The food corporations, such as Tyson's, Heinz and Monsanto have also participated in the illegal takeover of the federal government of the USA. They are guilty of illegally manipulating regulatory agencies, abusing campaign contribution laws, and exploiting lobbying to the point of treason. If you are wondering who owns the food corporations, it is the oil corporations; their evil is behind almost every aspect of corruption in America.

If Obama were to want to help America out, one thing he could do is making lobbying illegal. It was once a good idea, but has been perverted into a means for the ultra rich to bribe senators and that is not fair to the average person, who has no such influence over lawmakers.

The abuse of lobbying is simply treason, but the ultra rich have no moral conscience, nor any motivation outside of the acquisition of profit. They are happy to commit treason (like the destruction of the World Trade Center) in order to increase their bottom line. Treason means NOTHING to the ultra rich 1% of America. They consider themselves above the law. The law, after all, only exists to exploit the poor and make profit for the rich.

The federal government was originally created to govern over the USA. After the illegal takeover of the USA in the year 2000 by the traitor George W. Bush, the entire federal government was handed over to the corporations to do with as they wish. The Federal Government of the United States of America is now a corporate tool used for the management of their slave population.

Since the year 2000, Americans have had no federal government to help them in any way. The government is now just a façade that allows the Ultra Rich Corporate Lords to directly enslave the population of America, while brainwashing them into thinking that they have rights which do not actually exist.

Once upon a time, the president of the United States was supposed to pledge allegiance to the flag of the USA, to look out for the welfare of the people of the USA (not just the corporations), and to uphold the Constitution of the USA.

This is no longer the case. Well before the presidential election, each of the running candidates is forced to take an oath to maximize the profit of the Multinational Mega Corporations, the Ultra Rich Corporate Lords and the Oil Corporations that run American Society, even to the detriment of the Constitution and the People of the USA.

That's right; the corporations have made it impossible for the president to do his job.

We the people of the United States of America no longer have a federal government, nor do we have a president. The president is now more like a queen who has no real power, but holds the position in name alone. All three of the major branches of the government, as well as the regulatory commissions, are now owned lock, stock, and barrel by the URCL's and the ultra rich 1%, the aristocracy of America.

American culture had the potential to be morally good once upon a time, but it has degraded into a swirling cesspool of shit.

Although it is a little off the subject at this point, I now wish to discuss some of the aspects of modern culture that make me nauseated.

Chapter 3: General Issues

In addition to being amazingly corrupt and based on the evil of greed, modern day American culture has become so alien to me that it is no longer recognizable. I will start this chapter discussing some of the aspects of American culture in the year 2011 that offend me the most.

Soda Pop and Junk Food Society

The art of home cooking has been almost entirely lost. In spite of the popularity of the food networks and cooking shows, most Americans no nothing about cooking food.

When you go to the supermarket, look at the products other people are putting into their carts. A real cook will purchase raw ingredients from the vegetable, meat and dairy departments, and cook real food. An even more skilled cook will grow a garden, and produce as many ingredients as he can.

Retards, on the other hand, tend to go for the frozen food section, the junk food isles (chips, cereals, soda pop.) They fill their carts with cola, chips, frozen dinners, Twinkies, and ice cream. Such people don't have the first clue as to what to do in a kitchen. They are also eating food with little flavor and even less nutritional value. They can barely use a microwave oven, and frequently consume soda pop or "zombie juice" as I call it. Zombie Juice retards your brain's functions, so only stupid people drink pop, and as they do it, it makes them stupider.

Junk food is cheap, though. Some people buy the Kraft packaged macaroni and cheese dinners. That powdered "cheese" mix tastes like poison. The rest of it is just pasta, which you can get inexpensively if you buy it alone or in bulk. My approach to macaroni and cheese is to melt butter in a pan and thoroughly mix in a couple of tablespoons of flour. I then turn up the fire and slowly add milk or chicken broth until I have made a cream soup base. You can make many gravies and soups from that point, but for cheese sauce you grate in a cup or two of cheese, and when it is melted add a touch of salt and pepper. The last ingredient is just a dab of yellow mustard. You want enough to add a bit of tanginess, but not enough to actually taste or notice. It should be a subtle flavor. Anyway, the whole process takes no more time than making that packaged poison-tasting bullshit, and the result is macaroni and cheese that tastes amazing, and is probably healthier than that pre-packaged garbage.

I dare to call it laziness, but America has become a Junk Food/Fast Junk, Soda Pop addicted nation. I personally have not had zombie juice (ZJ) since 2005. I stopped drinking it because of the caffeine; I did not know at the time that it is laced with High Fructose Corn Poison (HFCP.) Even if I were to want to drink a soda pop, or offer one to a family member, I would make it at home by hand using organic sugar and natural ingredients. Anything that you make yourself is healthier and more delicious than a manufactured equivalent of the same product.

I am sick of seeing families who are all suffering from extreme carbohydrate poisoning loading up their carts with cases of Coke and Pepsi products. Even the Pepsi 0, which doesn't contain HFCP, does contain a large amount of aspartame, one of the most carcinogenic substances known. These fat, ugly, lazy fuckers are poisoning themselves to death, and it is an unpleasant sight to behold. I once met a girl who was nauseated every time she saw fat Iowans eating. I think that this was a little overboard, but I do get where she was coming from.

Jackasses

Here is another new development in American culture that I just don't get: the Jackass Movement. Of course, this is another good example of brainwashing. These people performing jackass stunts really are not jackasses at all. They are assholes. Being a jackass is often innocent in nature, saying something embarrassing in public for instance, or accidentally spilling a beverage on a stranger. Doing some dumb stunt on purpose fails to show innocence in nature. Instead, people willingly participate in stupidity, which makes them assholes and not jackasses. Most of these jackasses are also major egotists and total pricks as well.

Before I get off of the subject of jackasses, I want to make one comment about Tosh.o. I am not saying that his delivery is staccato and monotone, with strange gaps for a pre-recorded laugh-track. I am saying that his writers suck. It is a cute and clever idea to have a show based on what dicks people are on the internet. I just don't think that his writing staff is funny. I was watching Tosh.o one time and they showed a video of a cute little kitten. His comment was that he was glad that it made some woman feel less “lonely and childless”; I found this odd because I thought that the whole purpose of having children was to make people feel less lonely and cat-less.

I also want to point out that all reality TV is some form of jackass-ism. For one thing, it is NOT reality. Since there is a camera crew there, and a director, it is not reality. They use bad actors and scripts and choreography. There is nothing real about reality TV, and those who enjoy it are total idiots.

Of course, there are no egotists who are bigger assholes than jocks.

Jocks

America is obsessed with sports. They are a great distraction for those who can get into them. I have many objections to sports. I heard a friend defend sports saying that they are a great way for kids to learn about teamwork and cooperation. I on the other hand call that brainwashing. Learn to take orders so that some day you can be a good slave and not question the social system. Sports require people to work together and conform (the uniforms for instance.) Therefore, sports encourage the brainwashing and inherent evil in conformism.

We so worship our athletes in America that we are happy to forgive them for minor offenses such as murder (O.J) and rape (Coby.) Jocks are egotists who have their asses kissed their whole lives just because they demonstrate athletic prowess. I find competitiveness offensive, and also something that encourages conformism. Children are lead to believe that if they become sports stars, they will also be able to get away with unspeakable crimes. Jocks tend to be overly endowed with testosterone, which has a negative effect on the higher brain functions.

Human hormones are poison, essentially. Take the average jock as an example. This person tends to not excel in intellectual pursuits such as science and math, but does tend to "butt heads" and demonstrate violent behavior.

I have never seen a band member beating up on a member of the football team, but I have often witnessed several jocks kicking the crap out of some small kid who plays in the band. Any one of them could have kicked the kid's ass, but since jocks are conformists, they are also cowards. They just want to hurt someone who is helpless, not risk getting hurt themselves. Testosterone tends to lower the intelligence and raise the violence level of any individual. Taking steroids increases this effect. Men with lots of testosterone tend to be large, stupid and violent. Men with far less testosterone tend to show more aptitude in science and mathematics. It is too bad that most women go for the jock before the geek, because they are breeding stupidity into the American gene pool.

Estrogen, on the other hand, has a very different effect on an individual. Every man who has lived with a woman knows that she acts crazier when her hormone levels are fluctuating. I have even met women who act mature and poised up to about the age of 13, and can calm down and act sane after menopause. However, during the part of her life when she is fertile, she acts like a crazy person. This doesn't apply to all women, but I think that many men would agree that there is a hint of truth in this generalization. Estrogen, unlike testosterone, makes people emotionally unstable and mentally psychotic. Many women might take exception to the previous statement, but on some level deep down they must admit that to a degree, it is true.

We already have many parts of our culture that are an assault on our higher brain functions (like soda pop and advertising.) It doesn't help to have our bodies poisoned with hormones that make us even more stupid, violent, crazy and emotional. [Sigh!]

In my opinion, sports are inherently evil. Not only do they brainwash people with concepts like "teamwork" (total bullshit), but they also brainwash players into thinking that it okay for someone like a "coach" to tell them what to do. If you own your mind, it is NEVER okay for someone to tell you what to do.

In fact, I find everyone associated with sports evil and contemptible. I hate spectators and sports fans almost as much as I hate athletes and coaches. Sports are pure bullshit, and almost the entire country is brainwashed by them.

Materialism

Materialism is such an obvious vice of American society that it has been heavily criticized for the last half of a century. In spite of this, I feel obligated to discuss it, at least briefly.

Obviously, it seems shallow to obsess oneself with the acquisition of real estate, goods and technology. I think that it is more important to view the issue from the perspective of spirituality. Some people appear to be more obsessed with what they own than who they have in their lives. Others put emphasis upon the social aspects of their lives. I have found that it is a mistake to value objects over the living.

Stuff doesn't receive the emotional attachment that people do. If you really love someone, her happiness and presence should mean more to you than some piece of junk that they accidentally destroyed. Even when one of my kids trashes something, I don't get mad. I know that possessions can easily be replaced, however when a person passes away, the loss is permanent.

After all, your stuff is just a bunch of shit. More money can always buy you more stuff. I just get tired of being rejected by obsessed materialistic women who fail to get to know me on the inside. Instead they judge me upon the car I drive, the cell phone I carry, the watch, or the clothes I wear. I choose not to conform to the culture, so I don't wear a watch. I do own one, but I have not put it on since 2004. It is just sad when someone is so busy being obsessed with his shit that he loses the chance to say goodbye to a loved one before her death. Your stuff is replaceable; your loved ones are not. Don't fail to put your priority upon people, not upon your possessions. You will thank me some day if you can learn this one lesson.

Glamour Obsession

Being obsessed with glamour goes hand in hand with materialism. Americans somehow manage to value the shallowest of human concerns and desires. I can guarantee you that no matter how cute and perfect you are when you are young, you will either die, which would be tragic, or you will grow to be old as we all do.

I have seen women who grow old very gracefully, and remain pretty and even sexy well into their old age. If this is what nature gives you, then you are blessed. It is always better to age naturally than to go overboard with elective cosmetic surgeries.

I guess Michael Jackson would be one example of someone who got so addicted to plastic surgery that he eventually turned himself into a monster. On the other hand, this is appropriate because for one, I always hated the music of the Jackson Five because it was pop schlock. Further, Michael was mentally twisted and raped retarded boys. I think that being a monster on the outside is appropriate for someone who is a monster on the inside. Further, the child-fucker had no talent at all. His music was always shit.

Anyway, the term that I have coined that describes what far too many Hollywood celebrities do to themselves is "over-surgification". Some of these celebrity chicks have gotten themselves done over and over again until they all sort of look alike. I think you know the look that I am talking about. I call it the "overly-surgified" look. I understand the desire to look young, but let's not get so carried away that we turn ourselves into freaks, okay?

The Loss of Creativity

Sadly for the creative arts, it turns out that it is more profitable for the corporations to discourage creativity and actual talent. Using computers rather than people to generate forms of "art" is far less costly.

The music recording industry encourages forms of "music" like rap that require no musical talent at all. Forms of music that are no more than computer generated schlock are being forced upon the young and naive.

The film making industry is relying more and more on techniques that require voice talent and animation, rather than using real actors. Many movies rely on action and graphics to replace talent and good writing.

Paintings are now being produced on assembly lines instead of being created by real painters. The corporations are replacing computer automation with actual human creativity, and the arts are taking damage from which they will never recover.

My theory is that the corporations are now inclined to use computers to replace human input because it maximizes their bottom line. I hope that everyone can understand how this can lead to a degradation of true art and culture in general.

The Loss of Independent Thought

This argument goes directly to the most important issue: are you a slave or not? Anyone will claim to be capable of independence in one's own mind, but to what extent does this pervade into one's behavior?

In our society, it is easy to be lured in and overwhelmed by overt and subliminal brainwashing to the degree that one tends to screen it out consciously in order to maintain a normal life.

The attack upon an individual's consciousness in American society has reached unreasonable proportions.

The ultra rich aristocracy is comprised of white Christians, and they use the media to impose their own values upon the slave population of America.

If you think about it, we are bombarded with advertising everywhere we look in American culture. Make a list in your own mind as to how many different kinds and means of advertisement you are exposed to on a daily basis.

When I was young, there were far fewer forms of advertising. We didn't even have infomercials. There's an offensive concept: entire TV shows that are just long commercials. They now have networks that do nothing but advertise stuff.

Some shit-bag in the advertising industry got the idea of putting commercials on the screen DURING television shows! There are cartoons now that are based entirely on toy products for children, so the shows themselves are merely an advertisement for the toy line. Really good web sites don't use pop-up ads, or any form of advertisement at all that is unrelated to the main subject material of the site. As we all know, most web sites are over-commercialized and find it necessary to pollute the webscape with ads. All of this is in addition to the commercials and ads which were used back in the 1960's.

The advertising industry is choking the American public to death on advertisement!

A person is exposed to conservative propaganda in both advertising and the media which we view for enjoyment. Even TV shows, movies, the news and weather are crammed with brainwashing and the imposition of the values of the conservative faction of society. You literally cannot turn your head and view ANYTHING in America without someone trying to force their thoughts into your mind.

The Immoral Minority

White people (Anglo Saxon Caucasians) like to think that they are a majority in America. This is not the case. If you consider all of the people of African decent, all of the Hispanics, all of the people from the Middle East and all of the Asians who now live in the USA, there are clearly more of them than there are of Anglo Saxon immigrants from Europe. Anglo Saxons might be the largest minority in America, but they are a minority. White people can no longer claim to be the majority in America, a land stolen from the Native Americans.

Those who call themselves the "moral majority" also have no claims to morality. First off all, these conservatives are all

Christians, a faction long known for terrorizing the rest of the world, the Middle East in particular. Christians have practiced barbarianism, cannibalism, aggressive domination of other cultures, persecution of many peoples and religions, and mass murder of their OWN people! In fact, the white people whom I have been exposed to in the USA are morally the lowest slime that I have ever had the displeasure of coming into contact with. But then, most of that contact was in the context of their fists and cowboy boots smashing into my face and body. Of course I am going to hate white people, because of the hateful way they have treated me my entire life.

Those who call themselves the "moral majority" are the fact the most immoral people I know of. They are neither the majority nor are they moral, so I call the conservative white masses in the USA the "Immoral Minority."

Authoritarianism

As a non-conformist, I am and have always been anti-authoritarianism. I think that it is morally wrong to use any form of force or pressure to control another person's behavior. I hold this truth to be self-evident: the attempt to control another individual is simply evil.

Society, by its very nature, tries to control the individuals living within it. Many people would argue that this is necessary. They would argue that society could not function without laws, rules and guidelines. Even the word "society" implies the manipulation of the individual on a social level. The word "culture" is much more pure. A culture can be anything. The word society implies a social system, which requires the interaction of more than one human being. A hermit can have a culture of one, but he cannot claim to be a society without others to socialize with. To most people, the automatic assumption that society has the right to impose standards of conduct upon the individual is relatively innocent, and basically goes without saying.

In fact, conformists actually WANT to be told what to do, and given standards of conduct with which to conform. It actually frightens them to think of not being controlled! It also frightens them to think about unique and independent individuals who prefer to not be controlled by society, but then, conformists as we all know, are cowards.

It is my philosophy and practice to literally question everything. If society is required to control individuals' behavior using laws, does it not also restrict the freedom and liberty of the individual? When my freedom is restricted, I cannot have rights at the same time. Repressing my liberty makes me a slave, not a free person. Controlling my mind imprisons me by taking the right to choose my own action away from me. Society appears to take rights away, not grant them. And by what means does society restrict the behavior of individuals? How exactly does it take rights away? The answer is law enforcement.

I question the motivation of people who choose law enforcement as a vocation. The typical profile of such an individual is that of a schoolyard bully who takes pleasure in inflicting harm on smaller children. Most individuals who eventually end up with a career in law enforcement start when they are young torturing animals and tormenting other children. They grow up taking increasing amounts of pleasure inflicting harm on others, especially the helpless. Most law enforcement officers have killed at least one indigent person at some point in their careers. They take joy and pride in having a means of employment that allows them to use clubs, mace, hand cuffs, and firearms on fellow human beings. All police officers universally derive sexual pleasure from hurting people. In addition to that, most law enforcement officers participate in one or more illegal acts of corruption, including accepting bribes, dealing drugs, murder, rape, child molestation, illegal surveillance, behavior unbecoming a police officer, bribing Internal Affairs, kidnapping, and terrorism. They enjoy casual duties such as bodily cavity searches, putting people in bondage, and beating suspects. In short, law enforcement officers are monsters.

Society also practices very little control over the behavior of the police. I have seen many occasions when even after being convicted of murder, a cop will be briefly suspended and then allowed to continue work, rather than being fired and imprisoned as he should be. Cops get away with murder regularly, but if you object, they will murder you, so most people keep quiet.

Now how in the world could anyone derive enjoyment of any kind by inflicting pain upon another individual? In many cases, it is because the police officer was himself molested, beaten, and abused as a child. In other words, one theory is that monsters are made and not born. Stress inflicted by a painful environment causes some individuals to become sociopaths, and then enter the law enforcement vocation. Souls created by God find such things repugnant. If you are a good person, you should not ever enjoy the suffering of another person. Only evil enjoys such things. The definition of sadism is deriving enjoyment from the suffering of another person. It is pure evil. Individuals who enter the law enforcement profession have souls which were not created by God, but rather by Satan. This is how the police can experience enjoyment by hurting people: their souls were born out of Lucifer's pure evil.

Is this a bad thing? Most people will agree that society must have laws, and therefore we must also have law enforcement. So is it bad that law enforcement officers are sadistic evil cock-suckers?

Assume hypothetically that you are in charge of hiring police officers. Would you hire a pacifist who refuses to carry a weapon? NO. How about a decent, kind hearted individual who when making an arrest says, "Don't hurt him – let's be nice to him!" Is this someone who is truly qualified to be a cop? Not really. For law enforcement to be effective, it must be executed by sadistic, loathsome, evil individuals with no morals or personal restraint whatsoever. Society requires laws and law enforcement, therefore the evil of the police is also a necessity.

What if you were to tell the police that they can no longer restrain suspects, and that they are not allowed to use hand cuffs, clubs, or any kind of weapon? Would they not find it more difficult to do their job without their implements of sadism and evil? Clearly, for the police to do their jobs, they must be able to shoot people, beat people, and put people in bondage. These are acts of evil, but most people also consider them a necessity.

What then can we conclude about society, which by its nature destroys freedom and employs acts of evil? The obvious conclusion is that society itself is evil. Not just American society, and not just present day societies, but all societies throughout history and that are in existence today. All society, simply put, is evil.

Of course, law enforcement is not the only form of authoritarianism that we are forced to deal with in our society.

My aunt Esther is an authoritarian. She loves to boss people around! In fact, she actually prides herself on it. She is eighty years old now, and loves to tell stories about things she did when she was younger (anyone with an old aunt understands this experience.)

Unfortunately for many students in the past, Esther decided to become a teacher. She has many stories about her victories over children who were forced by law into the grammar school environment.

I hated school! I hate any hierarchical infrastructure, but I found the school system in the USA contemptible. Sure, I had some very good teachers, but I also knew the kind that my aunt was. These tiny, tin-plated, dictatorial, egotistical megalomaniacs just loved to torment the students. You know the kind; everyone has had at least one teacher like that in the past. Mrs. Bloom from Laurel School was like that. She actually taught well, but she was as mean as a snake! I actually made a paraffin effigy of her because I hated her so much. The fact is that far too many mean assholes are attracted to the education profession because they know that they will have a whole class of helpless students to lord it over.

My aunt Esther recently told me a story in which she was substituting for a teacher who was sick. She was called in to take over a class of eighth graders, and found one boy to be misbehaving. She said that she stood behind him and pressed down on his collar bone until he complained about the pain. Rather than deciding that he had learned his lesson, she told me that she stood behind this child and continued to inflict pain on him for the entire hour while she was presenting the lesson. This happened long before I was born, and it is probably not exactly what happened, but I was not there and can only take her word for it. The point is that she should be ashamed of having hurt an innocent child (if indeed she did hurt him at all.) She should be embarrassed and regretful, but instead she finds it a bragging point. She appears to be extremely proud of herself. I think that this is very evil behavior, and I can guarantee that a teacher could not get away with such an act these days. The child could just pull out his phone and call the department of child services or the police and have the teacher brought up on charges. If a teacher had done this to me, I would have removed my ruler from my desk and delivered a sharp jab to her trachea.

In another story she told me, she was walking down the sidewalk and she saw a little boy throw a candy wrapper on the ground. She recognized him as a boy who had behaved badly in school, and she spotted him immediately as a trouble maker! Fortunately for American Society, Esther was there to take quick action, and reprimanded the boy and forced him to pick up the candy wrapper. Yea! Esther to the rescue again! I sometimes wonder how we Americans could have survived the last eighty years were it not for the efforts of my aunt Esther! Sarcasm aside though, she really had no right to address this child in any way. If she had any authority over him, that authority was restricted to the classroom.

Further, this child's behavior was the responsibility of his parents, not her. She not only loves yelling at people, she is downright proud of it. Some people call this a "Napoleonic Complex"; I call it "Tiny Angry Person Syndrome" (TAPS.)

I am reminded of an incident a couple of years ago when my sister brought two bouquets of roses to my mother and my aunt on Valentine's Day. In Fairfield, this is considered a kind and polite gesture. Aunt Esther was furious! She not only made Dorothy remove the flowers from the house, but she also gave her a good piece of her mind by which to remember it! Dorothy didn't know that Esther is allergic to roses, and I am pretty certain that her act was not a premeditated attempt to cause harm. It was in fact, a totally innocent thing to do. Anyway, it did cure my sister of any bringing-over-flowers type behavior! Lesson learned: Esther prefers chocolates.

The point of all of this is that our educational system in America not only fails to educate by design and is full of propaganda and brainwashing, but it also has far too many asshole teachers with TAPS. Children should be educated at home on the internet. That way, they will be safer and they won't have to deal with asshole teachers.

Teachers have grown up in the USA being twisted with cultural propaganda and bullshit anyway. It is probably best to never allow a teacher near your kids because you don't know what they are going to fill their heads with. If you take charge of your kids' educations, then you have more control over the amount and kind of bullshit that is filling their heads.

All education is information from society, therefore it is all bullshit. I think that if people are going to raise kids, they should take the initiative to create the curriculum, and decide for them the base of bullshit that is going to brainwash them. After all, you don't want a stranger deciding what brainwashing and propaganda and bullshit are going to be fed into your children's minds, do you? You should be doing that yourself.

The central problem with authoritarianism is that it suppresses the freedom of the individual. Authoritarians are disturbed people who like telling others what to do. They generally have a very poor self image, and it makes them feel better to boss people around, because it boosts their egos.

What authoritarians need is therapy. If they could just talk out their childhood issues with someone and repair the self-stroking mechanism in their psyches, then they could act like well-adjusted members of society.

In American culture, we are brainwashed into accepting authority because it helps the URCL's to maintain control over their slave population. We are taught that authority is normal and we are taught not to question that for any reason at any time.

I question it because society has only proven itself evil. It has not proven to me that it is in some way non-malignant. I do not accept authority because I do not automatically accept that society is good, nor do I accept that it should have any authority over me at any time, for any reason. One reason that society is evil is that it is by definition bullshit. Any society or culture spinning some way to view the world is automatically bullshit. All society is bullshit; it always has been.

American Culture is Cancer

The most dangerous and sad things about our malignant, authoritarian, bullshit-based culture is that it is spreading all over the globe like some kind of cancer.

Take China, as one example. This is a country that has cultural tradition that goes back thousands of years. China once had a very beautiful culture. However, modern technology is changing all of that. Western Empire industrialized culture has a great appeal to those who live without the technology that we enjoy in the USA. The Chinese culture is desperately yearning for all kinds of modern technology, including cars, computers, and cellular phones.

Their ancient culture is being swept away quickly by Western Techno-culture. Temples are being razed and McDonald's being put up in their places. The old religions are dying, and the young are turning primarily to Christianity. China has made American English (NOT British English) their national language and it is now taught in all Chinese schools. The Chinese are hungry for all things American. They now have pop rock girl bands advertised on Coca Cola cans. They have literally given up an ancient and beautiful culture and adopted the cheap, commercialized rip-off of a culture of the USA.

The next thing you know, they'll have Chinese NASCAR, pro wrestling and line dancing!

The allure of technology is making Western culture desirable world-wide. Thousands of year's worth of tradition is being thrown away and being replaced with the toxic culture that we see around us every day in America. You can now buy Coke in every country in the world. They even have that poison on the Multi-national space station.

All of this feeds into the profit of the MMC's and the URCL's. Through their corporations, their illegal control over foreign and domestic political systems, and their advertisement campaigns, they are taking over the entire world. I don't know how to destroy the aristocracy, but its destruction is imperative.

Education

The American Education System, also known as the American System of Advanced Slave Training (ASAST) is bullshit. Instead of educating children, they brainwash them with propaganda.

Some Kindergarten teachers won't even tell the kids that Santa Claus and the Easter bunny are not real. They have to maintain the façade of bullshit, because that is what we do; we brainwash our children.

The internet is now the most powerful tool for education that has ever existed in human history. I suggest that we use it.

After all, the Chinese are using the internet, as are the Koreans and the Japanese. These countries all currently have better educational systems than the USA. They install internet connected terminal in the school buses so that the kids can study on the way to class, and it works. They have computers with internet access in all of the schools, and they graduate more genius level students every year than we have children total in the USA.

The countries that are improving their education systems using computers, the internet and common sense are now all above our standard of education in the USA.

And America just sits there acting like we just don't care that we are producing the first generation of totally brain-dead cybernetic zombies ever! Many American students who have computers in their classrooms use them to fuck around on You Tube and Facebook rather than look up material on Wikipedia that is relevant to the lesson being taught.

Even when we are using the right tools, the approach to education in America is wrong.

Obama has admitted that the lack of a primary education system in America is one of our country's biggest problems. I note that he has not done anything about it, but has some suggestions for patching the problem rather than fixing it.

As an engineer, I can tell you that there is a big difference between patching a defect in software and actually fixing it. With a patch, you just cover the problem over so that the customer doesn't know about it. A button that calls a function that crashes can just be removed, rather than fixing the function that is throwing the exception.

The problem with the American Education system is that it is broken; it does not work. It wasn't even designed with the intention of actual education being the result in the first place.

The solution that I would like to offer is that we sweep away the old broken system entirely. Private schools do a half-way decent job, so they can remain open, but the public schools must go. I suggest that education should be accomplished in the home.

In recent years, we have seen mass murders happening in schools. Have you heard of mass murders of home schooled children? No.

It is a terrible idea to get kids into classrooms. They are safer at home. You can implement security systems at home, and have a DVR record everything that your kids do. You can monitor their studies from work, and you can run a business from home so that you can watch over your kids as they study.

Some advantages are that your children's education will be better on the internet than it would be in a school, your children will have less exposure to drugs, violence, teenage sex, and communicable diseases. They will on average be healthier and better educated than children in school.

I also believe that virtual classrooms will allow children to benefit from the advantages of both situations. Teachers will also be safer if they are not in the same room as their pupils.

I don't know when Queen Obama is going to finally get to working on the education system, but I hardly see how he can make it any worse! I guess the pattern of executive behavior over the last decade suggests that the solution will somehow maximize profit for the MMC's and the aristocracy. Education itself will again be shoved off as some secondary concern. "Fixing" the education system will include more brainwashing, more propaganda, and more profit for the URCL's.

Health Care Reform

I hear people over-simplifying the problem with the American health care system. The problem is not that poor people cannot afford health care. That is a symptom of the problem, not the cause.

Before we can discuss a solution, we must first locate the source of the problem.

The real problem with the American health care system is the corruption of the oil corporations, and the industries that they use to victimize the American public. These victimizing industries include the pharmaceuticals industry and the medical insurance industry. Supporting these industries in any way makes the problem worse rather than better.

The medical insurance industry victimizes the poor. If one were to make insurance more available to the poor, all they are doing is making it easier for the poor to give what little money they have to the ultra rich aristocracy.

The WHOLE problem with Queen Obama's solution is that it is a patch, not a fix.

Obama's health care reform package attempts to make medical insurance more available to the poor. This fails to focus on the actual problem. The problem is that the medical insurance companies and the pharmaceuticals industry are victimizing the poor. If one doesn't fix this problem, then a solution will not follow. Giving the poor improved access to two industries that are victimizing them only makes the problem worse. As usual, the problem is the erosion of rights, not a lack of the availability of health care. Unless you are stupid, you should never give your money to a doctor, a pharmacist or an insurance company for any reason. They are your enemies.

The first important step in any intelligent health care reform is to get rid of the insurance industry. The Federal Government should take responsibility for the health of the American people and offer insurance to all citizens.

I note at this point that Obama is doing nothing about the Great American Food Crisis, but rather is allowing the food corporations to make everyone sick instead of caring for the health of the people.

One of the most important steps is to actually reduce the cost of health care. The medicines sold in America these days are artificially inflated in price by the pharmaceuticals industry. They do this to maximize their profit.

France has a comprehensive health care program that pays 100% of all medical costs for everyone in the county, whether they are a French citizen or just a tourist. Why can't America do the same thing? The answer is because the Corporate Lords won't allow it. In their minds, the slave population in America only exists to make them wealthy and powerful. We the people of America are treated like livestock.

When you begin to see yourself as the slave that you are, you realize that America is not about American citizens anymore. It is about American corporations.

My solution to the health care problem is for the government to offer free insurance to everyone, get rid of the medical insurance industry, force the pharmaceuticals industry to lower its prices, and have the government pay for 100% of the cost of health care.

It looks to me as though the Queen's health care reform package is only going to make the problem worse. It helps the corporations victimize the poor. It promotes the corruption rather than discouraging it. It fails to address any of the actual causes of the problem. Frankly, we would be better of with no health care at all than to have it forced upon us. Besides, Americans are too hypochondriacal anyway.

What needs to be done is educate Americans in how to deal with their own medical issues, and allow the individual to develop his own medicines and health care system. The medical industry kills more people annually than all other forms of death combined. People will be better off taking care of their own health.

Governmental Reformation

Most Americans are very unhappy with the way our Federal Government is handling matters these days. I have heard people talking about reforming the federal government, and others about replacing it altogether.

Obviously, all of the branches of the federal government are now so corrupt that they must be eliminated. The question then arises as to how to go about doing it.

Unlike most subjects, I have no solution to offer. Instead, I want to draw attention to the question as to when to reform the government.

Most people are so desperate for change that they would say that now is the time to act. They are correct; however there is a very important concern that must be addressed prior to any attempt at governmental reformation.

Before we discuss a solution, we must again locate the source of the problem. Sure, then Federal Government of the USA is totally and purely corrupt, but that is a symptom, not the cause of the problem.

The corruption that now holds our country's government in its grasp comes from the MMC's, the URCL's and the aristocracy. As long as the people who now bribe our congress are allowed to live, they will surely bribe our next Senate as well.

The only solution is one which the French implemented in the 18th century. Governmental reform that will make a difference can be implemented after we eliminate the source of the corruption.

There are many ways to take care of the ultra rich 1% ruling class in America. It is also important to keep in mind that we are talking about two to three million people, if you want to consider those rich monsters people, that is.

I prefer decapitation for the URCL's. It is a very certain form of execution. It is also dramatic, and people like it. I say we bring back using guillotines to deal with the ultra rich aristocracy and their families. Even so, any means of extinguishing their lives is acceptable. Hanging, poisoning, suffocating, shooting, burning, electrocuting and stabbing are all acceptable, as long as the goal is accomplished.

Once the URCL's and their families are dead, the people should seize their assets and use them to build a new government that represents and takes care of the people of America.

Chapter 4:
Technology - A danger to us all?

Technology Evolution

Modern technology is evolving fast; really, really fast! In fact, the effect is so pronounced and so obvious that anyone alive today can see how much technology has changed just within his own lifetime.

But it is not how much technology has changed in the last few decades, nor the amazing rate of the change that is as important as the fact that the rate of change itself is exponential.

This is a matter for concern for many reasons, and is the focus of this chapter. The exponential growth in the rate of the evolution of technology can and does cause problems for people on a personal level, on a cultural level, and on the physical level of the human DNA.

Technophobia

Most people who own a computer have a parent or grandparent who just doesn't get computers. Of course there are some old people who are quite skilled on a computer, but many won't even make the attempt to learn something new. I once told my mother to save an attached file from an email to her hard drive, and she literally screamed. "Save! Save! Save! I am tired of being told to save!" To me on the other hand, knowing the basics of file operations is just one of the things that one must know in order to use a computer. It goes without saying.

The generation of people over 80 in America these days generally have problems with electronic technology. I find it interesting that most children can operate a computer adeptly starting between the ages of two and five. The old generation struggled to learn to use a VCR, but when DVR's came out, they were lost. I gave my parents a TiVo for Christmas one year. My mom complained about it daily.

I went over to visit them and Ma complained that the God damned DVR was not working at all anymore! She picked up the remote and pressed the Pause button, and sure enough, nothing happened. This was no mystery to me at the time, because the DVR was on their coffee table, unplugged, and totally disconnected from the system. My parents however didn't understand why a component that is neither connected nor powered would fail to work. To them, it was just an appliance and should do its job without effort on their part. To me, it is a computer driven console device with a Linux operating system that is a component of their home theater entertainment system. Computers do not work without electricity. This is all obvious to me, but they were simply baffled.

The poor also frequently suffer from technophobia. The Schwans delivery dude is one good example.

Schwans now has a good on-line ordering system, but it used to have several defects. Whenever I would ask the guy coming to deliver the food about the web site, he would always comment that he didn't own a computer and doesn't have internet. I found this a bit shocking and out of stride with the current culture.

I met a cashier at the HyVee food store who actually seemed scared of the internet. She heard stories about identity theft, email hackers, and internet stalkers. She seemed to be the type who is intimidated by new technology in general.

I also met a gal who worked at Burger King who had never heard of High Fructose Corn Poison, so I told her that she could look it up on her favorite internet search engine. She explained that she doesn't have a computer or the internet. I got the impression that most of her friends and family were also not on line.

These are only a few examples, but I have noticed a disturbing trend in the Midwest for poor white people to not get more than a high school degree (in other words, just a dab of slave training) and are not on the internet at all. They have become a cultural cul-de-sac set outside of modern culture because of their lack of connectivity to technology. Most of them DO have cell phones, which is an interesting irony. Most hookers, pimps, and drug dealers also have cell phones these days.

I hold a Bachelors Degree in Computer Science. I have spent years developing computer systems for corporations, and I now own my own corporation, and continue to do development of computer systems. I am 47 years old (48 by the time you read this), and I have seen the first home computers and the birth of the internet. I have personally developed hardware, written firmware and software, and am a moderate system administrator, among many other computer skills. I also have an innate ability to analyze any technological system and repair it, regardless of whether it has computer components, or is more mechanical or biological in nature. Math, science, and technology are things I just "take to" naturally, and I have a tinkering skill that is extremely high. One would not think that I would suffer from an affliction such as technophobia. Technology is my freaking life!

And yet, when I look at the way technology is being used, I find myself offended by it. When I look at the way technology is being designed and implemented, I am frightened by it. It scares me even MORE that so many automatically accept new technology without questioning it or re-engineering it for their own particular use. Technology, like any aspect of culture, is dangerous and bullshit. In my opinion, people who walk blindly into technology without caution are just plain stupid. Of course, they are gullible and brainwashed, which is more to the point. They have been *conditioned* to act in an ignorant way without thinking about it.

It is in the interest of the corporations to program humans to automatically accept any technology they wish to shove down their throats.

Why is it a liability to accept modern technology? I will use the GPS as an example. It is awesome to have a device in your car that will help you navigate. However, it also allows one to be tracked by law enforcement and government officials. Many new vehicles are now being manufactured with this technology. On Star also incorporates GPS technology.

If fact, your cell phone has a GPS device too. Perhaps you don't *want* big brother watching you every move. Most people don't understand the details of how a GPS gets its power. You can't turn it off. The GPS in mobile technology is wired into the mobile device's main power harness. As long as there is *any* power in your device, the GPS will still work.

For example, if you turn off your cell phone, the GPS is still running. Many mobile devices have a backup battery to preserve memory while a user changes the main battery. If you remove the main battery of your cell phone, the GPS will run off of the backup battery, if there is one. The capacitors in the system will also hold a charge. Until the power system is totally drained (and all non-static memory is lost), the GPS will always run, and it cannot be turned off. Big Brother wants to keep an eye on you. If you own a cellular device and a new car, you probably have two GPS units on you at least part of the time.

One solution is to drive a car with no GPS installed, and buy or make a copper sleeve for your phone. Copper is used in labs to screen out signals. If you have a copper container for your mobile device, you can have privacy whenever you want it.

Keep in mind that the allure of new technology is great. It not only gives one more convenience, but also social status and egotistical gratification. Many people, without resistance, blindly accept any gadget that is shoved in their faces.

Technophilia

Technophilia takes many forms, and is not always just sexual arousal derived from contact with high forms of technology. Of course, there are people like that. Since the birth of the internet, we find that there are people who enjoy all manners of perversion. It's just that pre-internet they stayed more in the closet, where now everyone in the world can indulge in whatever gets them off.

The Super-geek is one form of over-obsession with technology. When I worked for Intermec, the engineers (geeks by definition) sometimes made fun of this guy we would see on the street. He had about 8 different devices strapped into holsters on his belt including different kinds of pagers, cellular devices, radios and portable palm-top computers. The guy was so into technology that he had to strap himself into all of that! This guy was so into technology that he looked like a freak even to other geeks! People with weak egos and low self-esteem often find a way to compensate by submerging themselves into some sort of cultural identity (like jocks.)

The Techno-egotist is another kind of human animal that has evolved in just the past couple of decades.

I remember when cellular phones had just been released into the public arena. The most obnoxious assholes would sometimes pull out their phone and just PRETEND to have a conversation in public just to look important to everyone around, like in a mall for instance. Whenever a new technology comes out, there are a few jackasses that are so desperate to be seen as an owner of that technology that they have to run out into public and stand around using it because they have no personality and their self-stroker doesn't work properly, so they have to fall back on having strangers stroke their egos for them.

I have seen teenagers going into public and texting just to get attention. They could just as well have stayed at home, but it makes them feel important to be seen doing it in public. It is a form of masturbation. I also want to mention that people who do this are nether knowledgeable nor do they posses the ability to learn. Techno-egotists are both retarded and lacking in character.

Have you ever not had access to your email and found it a very uncomfortable feeling? What if you made an eBay sale and you don't even know about it?

One time when my friends were helping me move into an apartment and as we were unpacking the moving van, we all made getting the internet up our first priority, so we could all check email. Every guy there felt uncomfortable until we got one desk put together and one computer assembled and on-line. Everyone checked their email, we all popped open a beer, and all was right with the world.

Have you had the experience that your internet is out, and you feel like you can't do anything? You might want to work outside, but you can't get the weather report. You have a business to attend, but you web server isn't on-line. You want to cook some food, but you can't check your recipes site.

All of a sudden it feels like your whole life has jerked to a sudden stop! Fixing the internet suddenly becomes your highest priority, and life won't feel right until you are on-line again. This sort of emotional response to the lack of the availability of technology I call Techno-addiction.

I have friends who can't even put down their God forsaken mobile technology, and that brings me to my next section, which is about Cyborgs.

Cyborgs

The technical definition of the word "cyborg" is a living being which is part biological and part technological in composition. It is a contraction for the words "cybernetic organism". These organisms were only to be found in science fictions when I was a kid, but now I see them everywhere.

If you are a living being whose life is totally dependant upon modern technology to survive, then you are a cyborg.

Most cyborgs, however, are in denial as to whether they are human or cyborg. It would therefore be advantageous at this time to offer a cyborg/human self-test that anyone can perform to find out if they might be a human being.

First, answer this question: Do you own a cell phone? If you don't have any kind of cellular network or internet connected mobile technology, or any artificial parts, then you are NOT a cyborg.

If you do own a cell phone, then move on to the next step. Take out your cell phone and turn it off. I do mean *all* the way off. Don't just suspend it or put in sleep mode, but turn it off completely. Place the cell phone in your sock drawer in your dresser in your bedroom, close the drawer, and walk away. I am now challenging you to not pick up or turn on your cell phone for any reason for seven days.

If you are a cyborg instead of a human being, you will immediately feel as though you are incomplete. Most cyborgs cannot even make it this far.

Most cyborgs are not able to turn off their cellular phones, not even for a minute. If you are like this and your life itself is totally dependant upon your mobile device, then you are not a human being, you are a cyborg.

The problem with cyborgs is that they are a menace to society. Their dependence upon technology poses a threat to the lives of anyone near them. It is also unbelievably rude when you see someone with one of those ear-phone cellular connections who walks through a store and checks out and doesn't even politely address the cashier because she is so entranced by her conversation. Also, have you noticed that a person having a conversation with one of those earpieces in has to talk louder so that everyone is forced to listen to him? It is egotistical and rude.

Personally, I think that it is creepy when you see someone doing this. On the other hand, people act like I am creepy when I walk around talking to myself in public. Perhaps *I* should get one of those earpieces, so they'll think I am having a cellular conversation. Maybe I should get some ear-piece mock-ups and distribute them to bums and winos.

When I took Drivers Education in high school, we were told to keep our hands on the wheel, our eyes on the road and our attention on traffic. We were also told to communicate with other drivers. I also want to add that there are laws in all fifty states that require drivers to keep *both* hands on the steering wheel while driving.

Cyborgs who attempt to operate non-hands free mobile technology and an automobile at the same time violate every *single* one of these rules and laws.

Traffic is obviously a life or death situation. People die in traffic accidents every day. Cyborgs that use their cell phone while driving are not just inconsiderate, they are criminals. They are increasing the likelihood that they are going to kill somebody's child, and they do it with an attitude as though they are somehow entitled to do it. After all, the cyborg's conversation and ego are more important than the lives of your family members, right? The police lie about it. Greg Francisco of the Fairfield Police Department told Ma that it is not illegal to use a cell phone while driving. This is not only a lie; it is illegal for him to say. There is an Iowan statute of the Department of Motor Vehicles that clearly states that it is not legal to operate a vehicle without both hands on the steering wheel. So the cops know that they are lying. They are just too lazy to actually DO the job they are paid for when they are not committing crimes against innocent people while on duty. A police officer lying about the law is clearly a matter for internal affairs, but cops are immoral by nature, or they would not be qualified for their positions as law enforcement officers.

I actually witnessed the following: A woman walked out of her house in Fairfield, IA while engaged in a conversation on her cellular phone. She walked over to her vehicle, and opened it with one hand, never lowering the phone from her ear. She then started up her car and pulled out into traffic, the whole time holding the cell phone up to her ear. She drove across town and finally arrived at Wal Mart, pulled into a parking space, turned off her car, and got out of her car while *still on the cell phone.* She accomplished the entire journey without ever having her attention completely on traffic. How can anyone be so irresponsible? Or inconsiderate? And note the brainwashing in the implication that it is alright for her to do so. Clearly, she was a criminal. If some kid had run out in front of her, she would not have been as prepared to deal with such an emergency situation until after the child had died under her wheels.

Cyborgs, because they risk everyone's lives including their own, are bullshit. Remember, bullshit can be anything disagreeable, immoral, or uncomfortable. It doesn't have to be something that is untrue. The fact that people actually do this is bullshit. It is obviously true; I see it happening all of the time.

Well, perhaps it is not acceptable for the state to impose law that would control cyborg behavior. Is there precedent for state law attempting to make the roads safer?

Yes. New York was among the first of the states to adopt a law that makes it illegal to use non-hands free technology while driving. Many states have followed suit, and recently Colorado also made it illegal to use a cell phone while driving.

So why is it that these states have adopted anti-cyborg laws, and Iowa has not? New York and Colorado both have reputations for having intelligent and enlightened laws. Iowa, on the other hand, is well known as the stupidest state in America.

Maybe Iowan lawmakers are just too fucking stupid to implement a law that would save lives. I can tell you now that Obama has no intention of introducing such a law on the national level, merely *because* it would save lives. It is curious how his agenda carefully avoids *anything* that would save the lives of American citizens. I guess it's just not important to the corporations who operate his marionette harness.

Cyborgs are not true human beings anyway. Therefore, it should not be considered murder if you have to dispense with one of them. Here is my trick: always shoot for the cybernetic appendage. If you can destroy the artificial technology, you can often salvage the remaining human/organic components. Unfortunately, cyborgs usually have technology attached to their ears, so discharging a weapon at this target frequently results in the destruction of the contents of the brain case. In the case of a cyborg, that organ is virtually unused, therefore worth a premium price on the organ donor market if you can acquire it intact.

Here is an example of one cyborg that I encountered who was certainly being a menace: I was headed south into town once and heading up a large hill I was trapped behind a menace asshole on a bicycle. This douche jock yuppie was on the cell phone, and weaving back and forth between the segments of the broken line in the middle of the road. There was a truck coming the other way that also could not pass this piece of shit, because he kept swinging back and forth through the breaks of the line, alternately occupying the right lane and the oncoming lane every time he crossed through the middle line. Could he have really been as oblivious to the traffic around him that he didn't realize that traffic could not go either way because of him? Or was he just enjoying seeing all of the cars trapped because he was an inconsiderate **dick**? On top of this, he was on the phone having a conversation and appeared to not have any idea that he deserved to be run down. If you see some piece of shit cyborg on a bike using a cell phone and causing problems to the drivers in traffic, steer your car towards them and accelerate. Tell the cops that it was an accident and that you didn't see him. It was his fault for being on the phone anyway.

I saw a woman drive straight though a stop-sign as though it was not there. I was not surprised when she drove in front of my car and I could see that she was on the cell phone.

I saw one guy drive right through a stop sign and make a left turn without signaling while on the cell phone. He then ran another stop sign wile making a right turn, again without signaling. In one block he had broken five laws, and these assholes act like they are entitled to act this way.

I want to point out that cellular phones are not evil. Not any more than guns are evil. What makes a cell phone dangerous is the asshole using it. After all, you have to get a license to own a gun, and you have to have a license to drive a vehicle. They should just force cell phone users to take a test to show that they know how to use one safely, and that they understand the law. Whenever a cop catches someone using the cell phone dangerously, they can then give them a ticket, and multiple offenders can have their cell licenses revoked, and their phones confiscated, and their services cancelled.

Zombies

Zombies experience life on a level removed from people who experience life as human beings.

For thousands of years, human beings have lived life while putting their attention on the world around them, on people and objects and forces in their immediate environment.

Modern technology has now provided a multitude of ways for people to put their attention on distant affairs, or the inter-workings of an environment that is a synthetic virtual simulation of reality. Many video games are based on virtual realities which the player interacts with via some configuration of control mechanism, all of which is controlled by the game's internal computer running software. Video games, which I have always loved, have now evolved to the point that they are now utterly mind-blowing! The dazzle of this kind of technology is irresistible to the young generation these days, and rightly so, it is by design.

Cell phones, computers and the internet allow people to interact with events, businesses and people at a distance.

We see that people now have the ability to put their attention on events not in their immediate surroundings. Telecommunications allow people to interact with a much broader amount of the world. Computers allow people to interact with worlds that don't actually exist at all. The trend I am trying to illustrate is that people often now live hours of their lives without even looking around at or focusing attention on their immediate surroundings. We spend so much of our lives in a virtual world that we lose touch with the real world around us.

In no case is this better illustrated than that of the Texting Zombie. Now, the texting asshole pile of shit zombie is not all bad. One of the funniest and entertaining things to do these days is watch texting zombies walk out in front of cars and make bets on whether they are going to die. If they do die – fuck them! The human gene pool cannot afford to suffer a zombie. If you do see one, hit the gas, not the breaks. Believe me, you are doing the zombie a favor too, not just society.

I actually saw a commercial recently in which fuck-bag texting zombies were walking into each other, stationary objects, and traffic. I am not sure what solution the ad was trying to offer, but it illustrates that I am not the only one who can see that this texting idea is really bad. We now have douche-bags who are putting no attention at all on the environment around them, even when they are in traffic.

I recently saw some shit-eater texting on some mobile device with his thumbs while he steered his truck with his elbows. He wasn't even looking at the road or at traffic. I find it disappointing to see someone who is that stupid, and yet not dead.

I later saw some idiot teenager guide his bicycle into the oncoming lane while steering with one hand and texting with the other, while taking a left turn through the busiest intersection in town. If the kid wanted to do something that obscene, he should have taken it to a private forum.

Just a few minutes later I had to break for some middle aged zombie who nearly walked in back of my car as I was attempting to back out of the parking space at the grocery store. It was a good thing that I saw him because this zombie turd was paying no attention at all to the traffic around him. He was also texting.

The next day I was crossing the street and had to walk out into highway traffic on 34 because this dumb bitch texting zombie was not paying attention to where she stopped her truck. She was risking pedestrians' lives because she didn't have her attention on her driving.

In just a couple of days I had seen totally retarded asshole zombies driving cars, walking in traffic and bicycling in traffic while texting. In every case, they were risking their own lives or someone else's. These ridiculous piles of shit are a serious menace to society and something must be done about them. When someone is so far gone on either drugs or technology that they are oblivious to the world around them, they are not experiencing normal human existence. Texting zombies cannot be considered to be human to any degree.

They should just put a bounty on them, and if you get a zombie, with your car for instance, then you should be able to call the game warden who should then pay you your bounty and remove the corpse. This will help human society in the short run, making the world a safer place. It will also get some of those retarded genes out of the population.

Besides, texting in traffic is always illegal in Iowa, regardless of whether you are a pedestrian, cyclist, or driver.

Technology Dependence

People frequently fail to think about the fact that almost everything they use or touch in a day was created with the assistance of computers. Just about every bite of food and every activity you experience is dependent upon computers in some way. Computers are now used for dating, food production, communications, transportation, recreation, monitoring the environment, shipping, manufacturing, measuring time, scientific research and masturbation. We have truly entered the age of computers.
As the average American lives his daily life, he rarely thinks about the fact that even his lunch was prepared by computers. People rarely think about the fact that modern cars all run on computers.

Most people would be utterly unprepared to deal with daily life if suddenly this technology was no longer available.

Try to envision a scenario where you suddenly have no support of computers in your life. You have nothing that you cannot create with your own hands. You have a few belongings, your hands and your mind;

those are your only assets. You don't even know exactly what time it is.

Now think about how you are going to live your life. Assume that there is no electricity or higher technology available. Without electricity, you have no computers.

Placing myself in the same scenario, I find that life would quickly become very boring. I would have no Netflix, no movies to watch, no internet and no video games. But I might survive where others will be wondering how to keep their asses alive.

I might have to use a file or manual grinding stone to sharpen my axes, but I would go to that trouble because I am going to need manual tools that are in good working order.

I could then spend a couple of hours cutting firewood in my forest before stoking up my wood burning stove. When the fire is going, I will have heat and a means of cooking food. I have lots of organic produce on hand, so I could cook up a soup for my lunch.

I have a scythe and a hand-crank grain grinder, so I can make flour and then either bake bread or make it into tortillas or pasta.

Eventually, it is going to get dark, so I know that I can read an old book by candle light. I make my own candles because I am also a beekeeper. I can start a batch of mead because it doesn't require much more than honey and water (and some yeast).

I might also dig myself a root cellar because it will help me keep food while I have no refrigerator working. I will also keep seeds in metal tins to keep the mice out of them. That way I will be able to grow crops and preserve food during the warm part of the year. I have a hand crank meat grinder too, and use it to make sausage from the hog I bought from a friend who lives nearby.

Even in a technology emergency, I will be able to provide food and drink and a warm shelter for myself and my family, because I own land and am prepared.

So what about the average douche who lives in a big city and has no skills or land or tools? That person is going to die.

If suddenly all computer technology and electricity were no longer available, society would turn into chaos in big cities.

Within the first few weeks, all stored food supplies would be exhausted. Within a few more weeks, all pets would be eaten. Within three months, starving people would be forced to turn to cannibalism in order to live.

The simple fact is that without crude oil, electricity and computers, the human race cannot possibly feed seven billion people. If modern technology suddenly disappeared, 90% to 98% of the human population would be dead within one year.

I am not arguing whether our race is or is not dependent upon modern technology to survive. I am saying that most people don't have any clue as to how dependent they are upon it.

Technology and Human Evolution

As discussed previously, technology has an effect upon human DNA and therefore our physiology. Technology is designed to make our lives easier and better. It removes pressures from the environment and compensates to make it easier for humans to survive in environments that would otherwise be far less hospitable.

The use of fire has cause the development of an omnivorous dentition, where our teeth used to more resemble a predator's dentition, with large canines and more differentiated kinds of teeth. The use of clothing has caused the loss of much of our body hair. Technology has always shaped it's users over time.

We now have technology that makes human survival easier than it has ever been in the past.

Many people can see some of the bigger dangers that technology can cause in the short run, such as nuclear wars, genetic manipulation, and biological warfare.

But these are by no means the most threatening dangers posed by technology. Consider, if you will, what the effect of such powerful technology will be in the long run. Humans have been given strong bodies to perform tasks which are being made obsolete by computer automation. Many of our skill sets which have allowed human survival in the past are now becoming obsolete. Humans are even losing the ability to digest real food and use their brains for higher cognitive functions, because these abilities are no longer necessary.

One of the reasons that the young generation is becoming a bunch of obese retards is because they can. Everything is done for them, and they are not required to do anything but eat Hot Pockets and plug themselves into a computer or a video game (or both.) Kids these days are losing the use of both their brains and their bodies because they need neither one in order to survive in today's world. If there is a major disaster, our young generation will be the first to die, completely unprepared to deal with any kind of adversity at all. Today's youth doesn't even posses developed skill sets. Most kids have no ability at all to deal with nature or survival.

Most of them don't even know how to grow a vegetable plant, or a grain crop. It's totally sad.

And yet we now have a techno-culture that favors genes that would otherwise be unfit to survive. Techno-culture now favors cyborgs and zombies and the kind of trash that the population's collective DNA pool would be better without.

Over time, technology, if it does its job correctly, will make human DNA less and less viable in the world's environment. At the same time, the world's environment is becoming more and more hostile. The more we become dependent upon technology, the less chance we have of surviving without it.

People who allow themselves to become totally dependant upon modern techno-culture without preparing for the fluctuations that any environment can experience over time are going to have no chance at all of surviving a major change in the planet.

Further, if we let our bodies and minds devolve to the point that we have no choice but technological dependency, then *any* threat to our technology would also threaten to wipe out the human species entirely.

Also, consider the exponential rate at which technology is evolving. Human culture evolves far less quickly than technology, and culture is already leaving some demographics behind because they cannot afford new modern technology. This is already causing divergence in the human gene pool in relation to economic stratification.

Human DNA evolves far less quickly than culture, which already cannot keep up with the evolution of technology. Technology will very soon and effortlessly make our existence on this planet totally obsolete. This will not necessarily mean that we will be extinct, just superfluous. When the human brain becomes obsolete, as we see demonstrated by the young generation, then humans will be completely retarded, and therefore worthless.

Surviving Disaster in Techno-Culture

It is not like it was ever really possible to separate technology from culture, but it has come to the point where modern high technology now defines culture. Culture itself is so computer and internet dependant that I now refer to it as “techno-culture”.

I think that the wise person in the 21st century will become a master of various kinds of technology. In the event of a true world-wide disaster, one should have a good inventory of hand tools in one’s possession.

In case the hand tools break down, one should have a skill set that allows them to produce and repair tools.

People these days think that electronic technology is the only technology. Most people don’t think of a hammer as technology. In fact, technology is defined as *any* tool that can be used for any purpose.

Humans have been developing various forms of technology for forty thousand years, at least. It serves the corporations for people to allow old technology to become obsolete and become dependent upon the new tech being offered.

However, I see a great deal of wisdom in maintaining the ability to repair and use various levels in the evolution of technology.

If your computer breaks down, your chain saw can still provide heat and fuel for cooking. If your chainsaw breaks down, you still can use your axe for the same purpose. If the axe breaks, I suggest you use a sharp stone.

The point being that it is the person in the future who is adept in all forms of technology who will be the most likely to survive, and who will be the most comfortable while they are alive.

If a person can just learn to make stone arrow heads, arrows and a bow by hand, and be able to hunt, they can have food, clothing, and the ability to defend a cave home, they can fall back upon Stone Age technology should it ever become necessary.

On the other hand, someone who is very adept in the use of hand tools might not be able to take advantage of high tech when presented with it.

The renaissance man of the 21st century will be able to program a computer, troubleshoot a system, or build a home from logs with an adz. Only the man who embraces all forms of technology will be fit to survive and raise offspring. Older technology is not going to go obsolete as the corporations wish for or expect. It is the ability of the engineer to embrace all technology of the past and create solutions for the present need that will give the adaptability that will be required for survival.

Older knowledge bases such as alchemy and the use of herbs for medication must be embraced and developed. But do you know how anxious the pharmaceuticals corporations are to see this happen? They discourage all more archaic and ancient forms of medicine because that is what maximizes their profit. They are willing to let people to suffer and die if it increases their bottom line.

I think that it is important to use the internet (while we have it) to amass as much information as possible on how things were done pre-industrial revolution, as well as have strong skills with modern technology, in order to have the best chance of dealing with whatever the future has to offer.

Chapter 5: Summary and Conclusion

America is currently faced with many problems and challenges. Instead of just illustrating one problem after another, I am attempting to offer possible solutions as well.

By now, it should be apparent to you that American culture is now more "full of bullshit" than ever. Also, culture by definition is bullshit; it is the very essence of bullshit. How then does one avoid being brainwashed by the propaganda and bullshit that comprises our culture?

Approach 1: Avoid it. Get a place far from other humans and have everything shipped in or delivered locally. Use any and all anti-advertising technology that you can access. Avoid as many forms of media as you can. Instead of reading a book, write one. Learn skills that allow you to be less dependent upon society. Learn to produce every food and product that you consume. Never put any "profile" on the internet for any reason.

Approach 2: Shield yourself from it. Practice the mental discipline of reminding your self that everything on the News is bullshit. Even if it is true, it is still a spun half-truth. Never automatically accept what humans have to say. Always question information, and ask yourself what is being left out, and what *motivation* the speaker has for saying it. Almost every word from a human mouth is intended to manipulate you in some way. Be very wary of that species.

Spam is a major problem in America, and it is getting worse. Look for anti-spam services that allow you to create a list of who can email you, and block everyone else. Anticipate the release of Schpleee Technologies Inc. Green List Technology-based Sender Management System.

Telemarketers victimize people, as well as do political campaigners, survey takers, charity callers, crank callers, and wrong numbers. Anticipate Schpleee Technologies Inc. Green List Technology-based Caller Management System. When the CMS is engaged, you will never be harassed by anyone who doesn't have your permission to call you.

Television advertising is one of the most powerful and evil sources of propaganda and brainwashing in history. New web-based services will soon be available that will allow one to enjoy TV programming from any device connected to the internet. Seek out services that detect and eliminate advertisements from the video stream. I plan to make such a service available.

Cyborgs, or "cell phone addicted assholes", are also a menace to society. My solution is to aim for the artificial appendage with the weapon of your choice. Attempt to salvage the biological components. Their brains are unused and of premium value.

Zombies, or texting assholes, are not human. Since they are unaware of their surroundings, they are particularly vulnerable. Use your creativity in eliminating them, or just enjoy watching them obliviously walk into traffic!

The American Education system, or System of Advanced Slave Training, is not designed to educate, but rather to brainwash. Keep your offspring home and educate them yourself using the internet; keep them safe with security systems.

Health Care Reform, if you really care about it at all, can be accomplished by eliminating the medical insurance industry, reducing the cost of healthcare, and having the Government pick up the tab for all costs.

The current reform package is only going to make things worse.

Aristocracy has been enslaving and murdering American citizens as though they are entitled to do so. They have obliterated democracy, and corrupted our Federal Government. I understand that many people are anxious for governmental reform, but I suggest that the corrupt aristocracy first be eliminated, so that the new government does not instantly become corrupt.

I would not even bother making suggestions as to how to make America better unless I feel that there are good things about America, and that it *is* worth salvaging.

Appendix A:
Good Things about America

I obviously don't want to offend anyone with my analysis of American culture. It would seem reasonable to consider what I think is good about the United States of America. This is, of course, totally subjective.

Creativity

Even the Maharishi Mahesh Yogi called the USA, "the most creative nation on Earth". Many nations have their own qualities to offer, superior production methods, better education, more manpower, high ambition, etc., but somehow America seems to be the best place for developing new ideas in technology. Maybe it is because we enjoy a high quality of life that allows us to focus on creativity rather than simply struggling to survive. At any rate, America has always offered creative solutions and this is one of the things that make it great.

Luxury and Comfort

As for the high standard of living, we have most comforts which humans usually desire readily available to us. As much as I complain about the fast foods, they would be considered a blessing in places where people are starving. We do have people who go hungry in America too, but generally we have no food shortage. We always have the most dazzling technology available to anyone who can afford it all of the time. If you have lots of money, you can buy some pretty amazing and bizarre shit these days! In America, it is great to be rich! We have every food imaginable available, and the best technology is also available all of the time, as long as you can afford it. In spite of the overpopulation problem, it is a great time to be alive! America is generally a really comfortable place to live. If you don't believe that, then notice how poor people (and even not-so-poor people) live in other countries. There are many people who would be grateful to receive what America's poor throw away.

Pets

We Americans even have so much excess and affluence that we can spend more on the care of pets than any culture has in history. America loves pets, and they are part of what makes life special. I think that the ability to truly love a non-human animal as family shows enlightenment. The beautiful pets who are part of our American families are one of the truly sweet things about America.

Women of Power: A New Paradigm

I recently watched a movie that was written in the early 1970's. Their portrayal of women was totally sad. The problem with the way 70's culture viewed women is that it was based on the 1950's stereotype. The women were written as being fairly dim intellectually and as helpless as children. This was awesome for a period of male chauvinism in our culture, but it is highly inappropriate in the view of 21st century culture.

If you look at the portrayal of women in today's media, it is a whole different story.

Look at the character of Lara Croft in Tomb Raider. Consider Milla Jovovich in The Fifth Element and in the Resident Evil series. Consider the women in The Seeker TV series. Look at Charmed and Buffy. Note how powerful Mira Sorvino's character is in The Last Templar. Kate Beckinsale in Van Helsing and Underworld, the list goes on and on. Even in Disney movies, women are now portrayed as powerful characters, not as helpless as children, as they have been in the past.

This view of powerful women is extremely rare in the history of human culture, and I am very happy to witness it in American culture. The fact that America now embraces the vision of women as powerful and equal to men is one of the things that make America a great place. The conservative brainwashing engine must have broken down somewhere along the line…

Natalietheism

To me, the greatest thing about America is that we have a living God, Natalie Portman. She is not the most special person in the world, because she is far more amazing than a person.

Whether you want to consider yourself a monotheist or a polytheist, we are lucky to be alive at the time when a living God walks among us. This only happens once every couple of thousand years. This is a blessed time to be alive.

If you do have internet, go to imdb.com and look up Your Highness. It is one of her new movies. She is also portrayed as a woman of great power; that's what I am talking about! She looks so hot that it drives me crazy. I love seeing her as an archer too, because I consider her Goddess of the Hunt, like Diana/Artemis.

Natalietheism as a religion was founded about five years ago and now has well over a million followers. The basic belief is that Natalie Portman created the entire universe, and is the source of everything, God herself.

When I see Natalie, I see God. There is no question about it; that is exactly what God looks like.

There are two major sects of Natalietheism: Nataliemonotheism and Nataliepolytheism. Nataliemonotheists simply believe that there is one true God and that is Natalie Portman. On the other hand, Nataliepolytheists are divided into two different categories. One believes that there are many gods, and that Natalie is the monarch that presides over all of them.

I personally am a polytheist and I believe in many gods. I believe in the God of the Sun, the God of the Moon, the God of the river, the God of the Harvest, etc., just as many polytheistic religions do. However, I believe that Natalie Portman is each and every one of those Gods.

Considering the amount of suffering and pain inflicted by the monotheistic God of the Western cultures, I see nothing wrong in believing in a God who has never inspired anyone to murder anyone else.

Appendix B:
More Stuff I *Hate!*

Bicyclists

I certainly am not saying that all bicyclists are assholes. There are many legitimate reasons to use bikes and many acceptable motivations as well.

For instance, when kids are too young to get a driver's license, it is acceptable for them to ride a bike, as long as they obey the traffic laws and motorists try to be wary of them. As long as no one is using mobile technology and no one gets hurt, it is fine.

I have total sympathy for the dude who is riding a bike and wearing a McDonald's or Burger King Uniform. Here is a guy who is willing to take a shit job in order to get the bills paid. If you just can't afford a car, it is fine to ride a bike, but would you at least kindly stop at the Stop signs please?

It's the spandex-wearing water-bottle toting, candy-assed looking helmet-wearing cocksucker who is just out there being an ass for no reason that pisses me off! They have bike trails that cars are not supposed to use, right? If they can go to all the trouble of buying all that crap just to ride a bike, they can also haul their asses to a bike trail and have a nice time without fucking up traffic! Egotistical motherfuckers!

RAGBRAI means “Real Assholes Going Bike Riding Across Iowa”. These piles of shit menace traffic all day, as normal highway and traffic routes are shut down for them. They then stop in some unfortunate town and get quite drunk before hitting the town like a crime wave. Take my word for it, if someone is a total asshole when they are on a bike, they almost certainly will continue to be an asshole after they have dismounted. I have seen it with my own eyes.

My first suggestion is to just stop the RAGBRAI event. My second is to issue the local townsfolk licenses to bag a couple of RAGBRAIers as they pass through town. That would put a pretty quick stop to RAGBRAI after a season or two.

Airlines

The problem with the airlines is that they shove stuff up your ass to beat you into submission. No matter how much they humiliate the passengers as they go through security, nothing is going to stop the airlines from entering into collaboration with the corporate lords to slaughter Americans if they feel like doing it again, as they did with the Bush syndicate. Their security is designed to manipulate and brainwash people who want to travel by air. Their entire security procedure, including making you remove your shoes, is oriented towards making humans feel like sheep. They actually treat you like you are an animal to impress upon you subconsciously that you ARE an animal, and therefore obviously have no rights. You might as well be going down a chute into a slaughterhouse and be electrocuted upon entry. Keep in mind that these bastards are *also* guilty of treason.

Laboratory Technicians

In millions of cancer research studies, it has been proven that the cause of cancer in laboratory rats *one hundred percent of the time* is laboratory technicians.

These technicians have been proven to administer known carcinogenic substances to laboratory rats topically, orally, and via inoculation.

Modern resolutions to this problem include security systems that prevent technicians from having access to laboratories and animal storage facilities, and also technician detection and elimination technologies.

It has been projected that by preventing the presence of laboratory technicians, the survival rate of laboratory rats will dramatically increase.

Obama's Policy on Corruption

I recently saw Queen Obama interviewed on TV. He was asked what he thought about lobbying and corruption. He acknowledged that it is a problem, and said, "There is nothing we can do about it."

I wonder what he means by "*we*". I think that what he meant was that there is nothing that he and his administration are planning to try to do about it.

Certainly, we American citizens can at least try to fix the sickness within our country.

After all, if we cannot have a corruption-free America, what is the point of having America at all?

Lingual Misdirection

If evil is something to be proud of, then the Bush administration should be commended for being excellent at it.

"Lingual Misdirection" is the art of making yourself less guilty by changing the terminology you use to describe it.

One example is that of "stealing the election." Well, if it is only stealing, then who fucking cares whether they are guilty? Hell, it is hardly even a crime! Stealing! Ha! I prefer to refer to subverting an executive election as treason. Using the proper word, "treason" carries the intended connotation that the offense was capital in nature. Further, there is no statute of limitations for treason, so I would like to see congress investigate and properly prosecute the Bush family for their many acts of treason against the USA.

The Bush administration was expert in using this kind of manipulation. The term, "Patriot Act" certainly *sounds* patriotic, but this was used to allow torture of innocent people, and destroy the Constitution of the United States. I guess they didn't think that the title, "The Treason Act" would have gotten many votes in congress. Congress knew that they were committing treason. This was not new to them; they also committed treason when they failed to prosecute the Supreme Court for letting a traitor take the oath of office.

Clearly, the Bushes as well and Cheney and Rumsfeld were guilty of massive terrorism against America (as well as Afghanistan and Iraq), yet they came up with the slogan, "War on Terror", even though *they are* the terrorists, and were attacking innocent people. Because of their evil behavior during the reign of Dictator Bush, I renamed his party the "Repugnant Party."

Even after Katrina, when they were arresting poor people, they kept saying that they were "saving" them. Why do you need a gun to "save" someone? They weren't saving anyone, they were illegally arresting them!

In fact, I think that everything that came out of the Bush administration was spun propaganda, brainwashing, and lies! This at least showed a true understanding of bullshit, and how to use it to manipulate the people of America.

Good old America! The land of the not so free and the brainwashed!